Mechanics of Ice Failure

Featuring real-world examples and practical methodology, this rigorous text explores time dependence in the mechanics of ice. Emphasizing the use of full scale data, and implementing risk-based design methods, mechanical theory is combined with design and modelling. Readers will gain understanding of fundamental concepts and modern advances of ice mechanics and ice failure processes, analysis of field data and use of probabilistic design methods, with applications to the interaction of ships and offshore structures with thick ice features or icebergs. This book highlights the use of viscoelastic theory, including nonlinearity with stress and the effects of microstructural change, in the mechanics of ice failure and fracture. The methods of design focus on risk analysis, with emphasis on rational limit-state principles and safety. Full discussion of historical discoveries and modern advances—including Hans Island, Molikpak and others—supports up-to-date methods and models to make this an ideal resource for designers and researchers.

Ian Jordaan is a consultant to C-CORE and Professor Emeritus at Memorial University, St John's, Canada. He is a Fellow of the Royal Society of Canada and has received the Horst Leipholz and Casimir Gzowski Medals, the P.L. Pratley Award, as well as several awards for best papers. He is the author of *Decisions under Uncertainty: Probabilistic Analysis for Engineering Decisions* (Cambridge University Press 2005).

Cambridge Ocean Technology Series

1. O. Faltinsen: *Sea Loads on Ships and Offshore Structures*
2. Roy Burcher and Louis J. Rydill: *Concepts in Submarine Design*
3. John P. Breslin and Poul Anderson: *Hydrodynamics of Ship Propellers*
4. R. A. Shenoi and J. F. Wellicome (eds): *Composite Materials in Maritime Structures Vol I*
5. R. A. Shenoi and J. F. Wellicome (eds): *Composite Materials in Maritime Structures Vol II*
6. Michel K. Ochi: *Ocean Waves: The Stochastic Approach*
7. Dong-Sheng Jeng: *Mechanics of Wave-Seabed-Structure Interactions: Modelling, Processes and Applications*
8. Johannes Falnes and Adi Kurniawan: *Ocean Waves and Oscillating Systems: Linear Interactions Including Wave-Energy Extraction*
9. Jeom Kee Paik: *Ship-Shaped Offshore Installations: Design, Construction, Operation, Healthcare and Decommissioning*
10. Bernard Molin: *Offshore Structure Hydrodynamics*
11. Ian Jordaan: *Mechanics of Ice Failure: An Engineering Analysis*

Mechanics of Ice Failure

An Engineering Analysis

IAN JORDAAN

Consultant, C-CORE, St John's, Canada

Shaftesbury Road, Cambridge CB2 8EA, United Kingdom

One Liberty Plaza, 20th Floor, New York, NY 10006, USA

477 Williamstown Road, Port Melbourne, VIC 3207, Australia

314–321, 3rd Floor, Plot 3, Splendor Forum, Jasola District Centre,
New Delhi – 110025, India

103 Penang Road, #05–06/07, Visioncrest Commercial, Singapore 238467

Cambridge University Press is part of Cambridge University Press & Assessment,
a department of the University of Cambridge.

We share the University's mission to contribute to society through the pursuit of
education, learning and research at the highest international levels of excellence.

www.cambridge.org
Information on this title: www.cambridge.org/9781108481601

DOI: 10.1017/9781108674454

First published 2023

Printed in the United Kingdom by TJ Books Limited, Padstow Cornwall

A catalogue record for this publication is available from the British Library

A Cataloging-in-Publication data record for this book is available from the Library of Congress

ISBN 978-1-108-48160-1 Hardback

Every second is of infinite value.
Johann Wolfgang von Goethe

'This book brings together many years of ongoing research Ian Jordaan and his teams conducted to elucidate the complex failure processes taking place inside an ice mass during its impact with an engineered structure. The model carefully factors in not a few but all observations stemming from well-instrumented field and laboratory experimentation.'

Paul D. Barrette, National Research Council of Canada

'*Mechanics of Ice Failure* is a must-read for both the beginner and educated in the field of ice mechanics. Ian Jordaan has dedicated a large portion of his life to a better understanding of the fundamentals of ice mechanics, and has spent countless hours composing his thoughts while preparing this book. It has been my honour and pleasure to have been a small part of Ian's journey!'

Paul Stuckey, Deputy Director, Ice and Ocean Engineering, C-CORE

Contents

Preface   *page* xi
Acknowledgements   xiii

Part I Basics and Design for Compressive Ice Interactions   **1**
1   **Introduction**   3
            1.1   Initial Comments: Elasticity and Dissipation   3
            1.2   The Ice Crystal   4
            1.3   Ice on the Earth   6

2   **Ice Behaviour under Stress and Tools for Analysis**   10
            2.1   Reality of Ice Failure   10
            2.2   Creep, Fracture, Strength and Damage: Ice Crushing   12
            2.3   Approach to Mechanics   14
            2.4   Creep of Ice   15
            2.5   Pressure Melting of Ice   18
            2.6   Dynamic Recrystallization   20
            2.7   Microstructurally Modified Ice: Development of Layers   21
            2.8   Thermally Activated Processes   24
            2.9   Fracture Processes   25

3   **Design Basics and Ship Trials**   30
            3.1   Local and Global Pressures: Exposure   30
            3.2   Data from Ship Trials   34
            3.3   The *Kigoriak* and the Alpha Method   34
            3.4   Initial Modelling of *hpz*s: CCGS *Louis S. St-Laurent* 1980   37
            3.5   The IDA Project: More Results for the Alpha Curve   40

4   **Fixed Structures and Medium-Scale Indentation Tests**   47
            4.1   Introduction   47
            4.2   Helsinki Tests, JOIA and STRICE   47
            4.3   Probabilistic Averaging   51
            4.4   Molikpaq   52
            4.5   Hans Island   57

4.6	Size Effect of Pressure and Ice Thickness	58
4.7	Medium Scale Indentation Tests	59
4.8	Ice-Induced Vibrations: Feedback between Ice and Structure	68
4.9	Field Tests on Iceberg Failure	70

5   Design for Ice Loading — **71**

5.1	Background and Lessons from Mechanics	71
5.2	Approach	72
5.3	Local Ice Pressures	73
5.4	Global Ice Pressures	75
5.5	The *Titanic*	79
5.6	Codes and Standards	80

Part II Theory of Time-Dependent Deformation and Associated Mechanics — **83**

6   Viscoelastic Theory and Ice Behaviour — **85**

6.1	Introductory Comments	85
6.2	Tension and Compression	87
6.3	Low Stress Regime of Springs and Dashpots: Maxwell Unit	87
6.4	Ice Crushed Layer as Viscous Fluid	91
6.5	Kelvin and Burgers Models	93
6.6	Hereditary Integrals	95
6.7	Retardation and Relaxation Spectra	98
6.8	Use of Nonlinear Dashpots: Spectrum for Andrade's Relationship βt^b	100
6.9	Stress and Strain Redistribution: Correspondence Principle	102
6.10	Early Damage Mechanics and Experiments	103
6.11	Schapery's Use of Modified Superposition Principle (MSP) in Ice Mechanics	108

7   Complex States of Stress and Triaxial Tests — **112**

7.1	Multiaxial States of Stress	112
7.2	Review of Past Triaxial Testing	115
7.3	Initial Finite Element and Damage Analyses	117
7.4	Targeted Triaxial Tests	119
7.5	Barrette's Analysis	128
7.6	Localization of Damage: Runaways in Triaxial Testing	130

8   Damage Analysis and Layer Formation — **135**

8.1	Damage Analysis Accounting for Pressure Softening	135
8.2	Simplification: Incompressible Power-Law Materials	146
8.3	Crushed Ice Extrusion Tests	148
8.4	Singh's Analysis of Crushed Ice	151
8.5	Thermal Matters	154
8.6	Small Scale Tests: Synchronization	156
8.7	Scaling in Ice Mechanics	164

9   Fracture of Ice and Its Time-Dependency   169

9.1   Time-Dependent Fracture in Ice–Structure Interaction   169
9.2   Measurements of Fracture Toughness   170
9.3   Introductory Theory and the Barenblatt Crack   171
9.4   Schapery's Linear Viscoelastic Fracture Theory   177
9.5   Nonlinear Viscoelastic Fracture Theory   182
9.6   Experimental Comparisons: Beam Bending and Compact Tension
       Experiments   187

10   Concluding Remarks   190

Appendix A Précis of the Work of R. A. Schapery   192

A.1   R. A. Schapery   192
A.2   Introductory Comments   192
A.3   Biot–Schapery Theory   193
A.4   Constitutive Theory Using Modified Superposition Principle (MSP)   199
A.5   Consideration of Damage   202

Appendix B Preparation of Laboratory Ice Test Samples   208

B.1   Triaxial Test Specimens   208
B.2   Laboratory Indentation Tests   209

References   211
Index   227

Preface

The present work results from many years of research into ice–structure interaction, mainly aimed at safe design of engineered structures. A concentrated effort commenced when I joined DNV Canada's Cold Climate Technology group in Calgary in the 1980s. I then joined the Faculty of Engineering and Applied Science at Memorial University in St John's in 1986, taking up a research professorship sponsored by NSERC and Mobil Oil Canada. I continued involvement in engineering practice, designing offshore structures to resist ice actions. At the same time, it was possible to develop a programme of research into the mechanics of ice.

My experience in practical design includes work on developments offshore Newfoundland, including jackup rigs, gravity based structures, FPSOs and shuttle tankers. I have also worked on concepts for arctic development in the Chukchi, Beaufort and Barents Seas, and for the Caspian Sea, as well as the design of the Confederation Bridge in Canada. I chaired a review of the Proposed Revisions to the Canadian Arctic Shipping Pollution Prevention Regulations and worked on problems of arctic shipping. The work included an extensive review of data from ship trials, especially the Polar Sea data sets. Data from the Molikpaq arctic project were studied, leading to a re-evaluation of ice load estimates.

During 2003, I was asked to join an expert panel of the Royal Society of Canada to study science issues related to oil and gas development in the offshore regions of British Columbia. The work was carried out for the Minister of Natural Resources, Government of Canada.

The book describes directions I have taken in dealing with the use of data and in implementing risk-based design methods. The use of mechanics in analysing ice–structure interaction has generally not reached a level that is acceptable for direct use in design, so the book is divided into two parts, the first dealing with basics of ice mechanics and approaches to design and the second dealing with mechanics in more detail. The approach in Part I of the book includes an extensive analysis of field data and development of probabilistic design methods, but at the same time recognizes the necessity of obtaining a fundamental understanding of ice failure processes. In Part II, dealing with ice mechanics, the focus is the use of viscoelastic theory with damage in dealing with ice crushing, compressive failure of ice and fracture. Generally the applications are concerned with interaction of ships and offshore structures with thick ice features or icebergs. Along with the work on design for ice loads, a dedicated research effort has been conducted into ice failure, including medium scale field tests and their analogue in the laboratory at small scale, but with scaling of the mechanics.

The main overarching theme of the work is the importance of time in the mechanics of ice, in particular on the mechanical behaviour of ice. Not only does ice creep under stress, but its microstructure alters under stress with the occurrence of extensive micro-cracking and grain refinement. These effects, in turn, lead to profound changes to the mechanical response, with the viscoelastic compliance increasing by many orders of magnitude under certain states of stress. The book details a personal account of experience with ice mechanics, a journey with some discoveries along the way. The book does not claim in any way to be a complete review of ice mechanics in the literature. The main point of emphasis in mechanics is the importance of time in understanding ice response to stress. Many researchers have neglected this aspect, and in some cases have used elasticity- or plasticity-based theories in which time does not appear. These cannot come near to solving the problems of ice failure, especially the failure in compression. In these approaches, tests are made on small scale specimens and a failure criterion might be deduced. The results might then be incorporated into a continuum analysis where the failure is assumed to occur locally corresponding to the failure criterion deduced, but in reality, all-important microstructural details such as microfracturing and recrystallization need to be included. These have to be studied using triaxial tests and by the use of damage mechanics coupled with viscoelasticity. The state of stress proved to be an all-important aspect. Attempts to take into account the state of stress using time-independent strength parameters, for instance by using pressure-dependent yield criteria, cannot deal with the reality of the time-dependent history of microstructural change.

It was found that the work of R. A. Schapery on viscoelastic theory, fracture and damage provides unique insights into analysis so this is given special attention, including an appendix (Appendix A) that contains a summary of the relevant theory. The final chapters of the book contain summaries of theory that might prove useful in future research. It is hoped that this will encourage further work in this area and the use of viscoelastic theory and damage mechanics.

Acknowledgements

I was fortunate enough to meet Ken Croasdale soon after I commenced work on arctic engineering, and he responded generously with his great experience to my questions. He always had an incisive answer, having thought through the basic issues, and had many modelling ideas and idealizations. We have worked together on several projects, including studies for development of the Shtokman natural gas field and the Caspian Sea oil projects. I also developed a close association with the National Research Council of Canada, with Garry Timco, and with Bob Frederking. After the Pond Inlet tests in 1984, the apparatus was donated to Memorial University by Mobil Oil Canada Properties and Bob and I developed two new field research programmes as well as tests on crushed ice using the apparatus. This was done in collaboration with Dan Masterson and his group at Sandwell Inc., who were very supportive. Bob was an excellent collaborator, and he led the field programmes with great skill and dedication. He provided every assistance in the preparation of the present work, including provision of copies of photographs taken by him during the Hobson's Choice field indentation experiments.

Much work has been carried out in collaboration with industry partners. I mention Phil Clark, formerly of Husky Energy (now Cenovus Energy), and Adel Younan of ExxonMobil, Pavel Liferov and Arne Gürtner of Equinor, John Fitzpatrick, Kevin Hewitt and Peter Noble, who have worked in many offshore projects in the Canadian offshore. John's description of the Hans Island experiments, in which he was a participant, is very informative and summarized in the book. There was a very significant flowering of understanding of ice mechanics in Canada in the 1970s and 1980s, during which major discoveries regarding high-pressure zones and ice pressures, both local and global, were made. I had the pleasure and honour to work with Michel Metge and the late Dan Masterson in projects related to the advances in knowledge. A study of engineering in Canada's north is given in Croasdale et al. (2016).

I mention especially my collaboration with C-CORE in St John's. Soon after joining Memorial University, I formed an association with this consulting organization. They have specialized in questions relating to offshore engineering in harsh environments, and especially the question of ice–structure interaction. Many industry-based projects have been completed in collaboration with C-CORE. Judith Whittick and Charles Randell, both formerly serving as President and CEO, have always been supportive. I have worked on many projects with Freeman Ralph, Vice President, and have much appreciated this collaboration. He has been consistently supportive of the present endeavour.

Very special thanks are noted for Paul Stuckey who assisted in every way, helping with proof reading and in the preparation of the figures and illustrations, and in giving advice in general. I am immensely grateful for his careful work and assistance. In figure preparation, we were expertly assisted by Zarif bin Feroze. I also appreciate the collaborative work with Mark Fuglem, Jonathon Bruce, John Barrett and Tony King of C-CORE.

Richard McKenna assisted very effectively in early modelling of damage mechanics using finite element analysis; the laboratory effort was expertly started by Barry Stone. Irene Meglis carried this on, testing at higher confining pressures, producing wonderful thin sections and showing great care to detail. Dmitri Matskevitch helped to further develop the modelling and finite element programming. Jing Xiao applied great skills in formulating the finite element solution.

I thank especially my colleague Paul Barrette, now with National Research Council Canada, who has contributed significantly to the approach taken to ice failure and who has conducted unique experiments of his own including high-pressure triaxial tests. He initiated very ably the experimental work on small scale specimens. Jennifer Wells continued the work, using high-speed video and pressure-sensitive film.

I thank sincerely past research staff and graduate students at Memorial University: Shawn Kenny, David Finn, Rick Meaney, Kurt Kennedy, Peter Brown, Richard McKenna, Jing Xiao, Bin Zou, Bin Liu, Michelle Johnston, Barry Stone, Irene Meglis, Dmitri Matskevitch, Paul Melanson, Trevor Butler, Karen Muggeridge, Thomas Mackey, Jennifer Wells, Sanjay Singh, Chuanke Li, Denise Sudom, Jonathon Bruce, Rocky Taylor, Thomas Browne, Joshua Turner, Mark Kavanagh and Brian O'Rourke. Rocky Taylor assisted in the development and execution of several projects. The ingenuity and dedication of the persons noted made the present work possible.

Part I

Basics and Design for Compressive Ice Interactions

1 Introduction

1.1 Initial Comments: Elasticity and Dissipation

Ice is present in many offshore locations where engineering activities are required. These activities include transportation and movement of materials related to human habitation, as well as resource development and use. The facilities for these activities require consideration of ice loadings and associated design. Transportation involves design of ice-strengthened vessels and icebreakers. Ice is a complex material, existing on earth at temperatures near its melting point. It is prone to creep under stress and to fracture. Yet it can produce extremely high local stresses on icebreaking vessels and structures, and ice can interact with fixed structures causing significant vibrations, with energy flowing into and out of the structure. The focus in the present work is interaction of structures with thick ice features and icebergs.

The stretching of a spring which returns to its original position after release of the stretching force embodies the idea of elasticity. In contrast to the elastic storing of energy, most mechanical processes involve dissipation and the creation of heat. The dissipation can occur internally, within the material, for instance the glide motion of dislocations in materials such as steel and ice, thereby generating heat, a process of internal friction. Dissipation can arise also from changes to the microstructure of the material being considered. Ice is a case in point, with large changes to crystal size involving dynamic recrystallization under high confinement and shear. The consequent changes to mechanical response are extreme and offer explanations for failure processes. We aim at a fundamental approach to mechanics of materials, in particular time-dependency, which accounts for such factors. The ideas of storage and dissipation of energy in mechanics can be represented by springs and dashpots, respectively (Figure 1.1), and these provide an interpretation of storage (essentially time-independent deformation in a spring), and dissipation (time-dependent dashpot movement). The elastic modulus is E, relating stress and strain, with viscous movement (or "flow") governed by the viscosity μ, relating stress and strain rate. The elements can be linear or nonlinear with stress. If linear, the moduli are constant and the units are Pa and Pa·s respectively.

Ice exhibits a wide range of material responses and as a result is a very interesting material to study. It is particularly prone to viscoelastic movements, with a high degree of time-dependence, as well as to fracture. It is prone also to change its microstructure under stress, and the resulting effect on its mechanical response is highly significant.

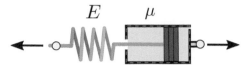

Figure 1.1 Spring and dashpot: basic mechanical elements for material modelling.

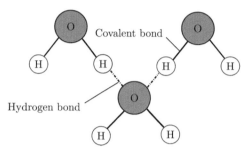

Figure 1.2 Bonds in ice crystal.

Its mechanics serve as an extreme which provides an excellent basis for studying and understanding material response. Despite the complexity of ice behaviour, engineers must devise ways to design engineering structures and vessels to resist ice forces. The approach taken in this work is to rely on a combination of basic knowledge and field measurements, with empirical interpretation of data and measurements.

A significant amount of engineering research took place in Canada during the 1970s and 1980s. This included a series of highly imaginative experiments such as the use of instrumented icebreakers in ramming of hard ice features and iceberg bergy bits, the Hans Island experiments, the Molikpaq instrumented platform in the Beaufort Sea, and experiments involving medium scale indentation of ice in the field. These experiments led to an important increase in knowledge as reported in the present work.

1.2   The Ice Crystal

Bonds will cause atoms to combine into molecules and thereby into liquids and solids. The main forms of bond between atoms are ionic, covalent and metallic. A covalent bond is formed when the valence electrons from one atom are shared between two or more atoms. The water molecule H_2O consists of two covalent bonds between the oxygen atom and each of the two hydrogen atoms. In water and ice there are much weaker hydrogen bonds between the H_2O molecules themselves, and the number of these decreases with increasing temperature of water. Figure 1.2 illustrates the two kinds of bonds. Liquid water molecules are in constant motion and will have on average fewer than 4 (generally between 3 and 4) possible hydrogen bonds. In ice, water will form all 4 possible hydrogen bonds because the molecules in a solid are locked in a fixed crystal lattice. As water is heated, the heat added contributes to loosening, distorting or breaking the hydrogen bonds. About 10% of the hydrogen bonds break upon melting.

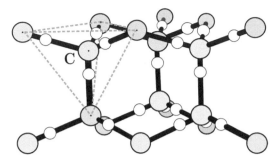

Figure 1.3 Ice crystal structure: basic unit showing positions of atoms.

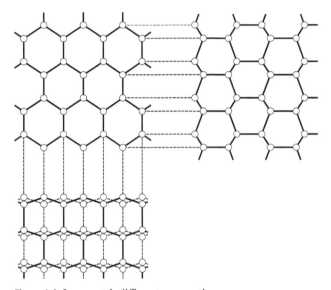

Figure 1.4 Ice crystal: different perspectives.

Ice Ih, the familiar "ordinary ice" is the hexagonal form found on earth. (Other forms do exist, for instance ice Ic under high pressure; see, for example Schulson and Duval, 2009. These are not relevant to the present work.) Figure 1.3 illustrates the arrangement of the oxygen and hydrogen atoms into a hexagonal structure, with Figure 1.4 showing the arrangement looking from different directions. A useful representation is obtained by considering a particular oxygen atom, say "C" in Figure 1.3, and its four nearest neighbours, which are shown by broken lines that are joined to the atoms surrounding "C" in the figure. This very closely forms a tetrahedron with C at its centre. An excellent first approximation to the geometry is obtained by considering the oxygen atoms as being arranged in a set of regular tetrahedra; see also Pounder (1965) and Schulson (1999). Some tetrahedra will point down as shown in Figure 1.3; others will point upwards. Figure 1.4 shows different perspectives of the ice crystal.

The c-axis of the crystal is along the axis of the hexagon while the basal plane is at right angles to this axis, as shown in Figure 1.5. To emphasize, the oxygen atom of each molecule is strongly bonded to two hydrogen atoms by covalent bonds, while the molecules are weakly bonded to each other by hydrogen bonds. The hydrogen atoms

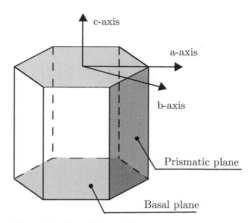

Figure 1.5 Crystallographic planes and axes.

do not lie midway between two oxygen atoms, but are closer, and more bonded, to one of the two. This leads to uncertainty and an increase in entropy. The distance between two adjacent oxygen atoms is 0.275 nm.

Associated with the anisotropy is the property of double refraction and birefringence. As a result of this, interesting details can be found regarding the microstructure using thin sections of ice samples viewed through crossed polaroids. These include the crystal structure, microcracking and recrystallization, all very important in the present work. Details of this technique can be found in Pounder (1965) applied to ice and more generally in mineralogy textbooks. The use of the universal stage to obtain crystal orientation is noted. The research carried out by the author and reported in this work generally followed as far as possible the demanding methods of Sinha (1977) in preparing thin sections and studying crystal structure. Sinha also pioneered the use of side lighting across the thin sections to identify cracking and microcracking. These tools were instrumental in discovering the effect of stress history on microstructure under various states of stress. Examples of thin sections can be found in Chapter 7 (Section 7.4) and in colour in Melanson et al. (1999) and especially in Meglis et al. (1999).

1.3   Ice on the Earth

Ice exists on earth at high homologous temperature. For ice at −10°C (263 K), the homologous temperature T_m is 263/273 = 0.96. The fact that water expands upon freezing results in masses of ice floating on water. Both water and ice have unusually high values of specific heat capacity, as compared to soil and rock, taking much longer to warm in summer, and to cool down in winter. As a result, water and ice provide a thermal stabilizing effect on the climate. Milder temperatures result in maritime regions. In the arctic and antarctic cryosphere, cold temperatures persist, aided by the high albedo associated with the snow and ice cover. The floating ice at the same time presents a hazard—and a challenge—to shipping and to construction in offshore regions.

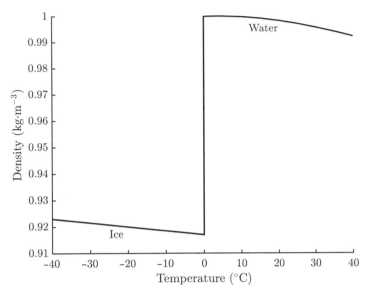

Figure 1.6 Density of ice and water at atmospheric pressure.

We give here an overview of ice on earth. Detailed descriptions can be found for example in Pounder (1965), Sanderson (1988), and Weeks (2010). As noted in the preceding section, ice on earth is in the form of hexagonal crystals. We consider first freshwater ice and then sea ice, starting with the freezing process of pure ice and water. The density of ice and water varies considerably with temperature as shown in Figure 1.6. Pure water expands about 9% upon freezing, a large amount even compared to other materials that expand upon freezing. This can cause huge pressures to develop in sealed containers of water that are then frozen. Figure 1.7 shows the density variation of water near the freezing point. It is seen that the maximum density occurs at 4°C, termed the inversion temperature. At this temperature, clusters of crystalline ice form, which have lower density than the surrounding water molecules, thus resulting in a decrease of density with temperatures below 4°C. When a body of water at a temperature greater than 4°C is cooled by the air at a lower temperature, the cooled water is more dense and sinks. This vertical circulation continues until the water reaches a uniform temperature of 4°C. If the water is cooled further, the density is reduced so that the vertical (convective) circulation ceases. Further heat transfer is by conduction, a much slower process. Ice will then form on the surface, and ice cover will develop. The temperature at the bottom of the ice will be 0°C, and the water will have a gradient from 0°C to 4°C, termed the thermocline, below which the water is at 4°C. Wind, wave and current will modify and complicate this idealized picture.

The presence of salt in water modifies the situation considerably. The salinity of sea water is relatively constant at 35 parts per thousand, denoted 35‰. (There are some exceptions such as the Baltic and Caspian Seas, which are brackish.) Both the temperature of maximum density and the freezing point decrease with salinity, the first at a greater rate, so that at salinities greater than about 25‰, sea water has maximum density at the freezing temperature. There is no temperature inversion and cooling of

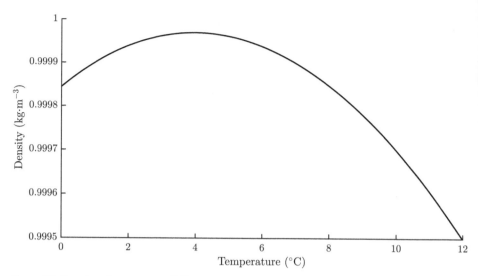

Figure 1.7 Density of water near 4°C.

the ocean surface by a cold atmosphere will result in the surface water being more dense. As a result, convection will continue down to the freezing point and the whole body of water must be reduced to freezing point in order to form ice. Ice cover does not form as in freshwater lakes. Only under very cold conditions with protracted periods of low temperature, such as at high latitudes, will ice form on the sea. This is well known in arctic conditions, which will now be discussed.

The term "young ice" refers to ice that is 10 to 30 cm in thickness, formed recently.[1] First year ice is floating ice of no more than one year's growth, developing from young ice; thickness is from 0.3 to 2 m. It is usually level where undisturbed by lateral pressure. Currents and winds cause the ice floes to interact with each other, resulting in pressure ridge formation. The ridges are made up of angular ice blocks of various sizes that pile up onto the floes. The part above the water level is the sail and that under the water level is the keel. Grounded ridges are termed stamukhi. First year ice features that are ridged can be rough and irregular, although linear in plan. Second year sea ice has not melted in its first summer. By the end of the second winter, the thickness might be 2 m or more. Summer melting results in smoothed and rounded ridges and hummocks.

First year (FY) sea ice generally has a columnar structure (Pounder, 1965; Sanderson, 1988). In the initial freezing on the water surface, small platelets of ice form, generally less than a millimetre in thickness and a few centimetres in width. This is termed frazil, or frazil ice. The c-axes are aligned at right angles to the plane of the platelets, and are generally pointed vertically. Large numbers of these form with c-axes parallel and pointing vertically. These act as seed crystals for the continued growth of ice. The crystals, or grains, which grow perpendicular to the c-axis grow more rapidly

[1] With regard to nomenclature, there are various national standards. The WMO Sea Ice Nomenclature is one useful source.

than those which grow in the direction of the c-axis and predominate in the further growth of the ice sheet. The result is that columns of ice form with c-axes pointing horizontally, so-called columnar ice. This ice is referred to as S2 ice if the c-axes are randomly oriented but in the horizontal plane. S1 ice has vertically oriented c-axes, and S3 denotes ice with the c-axes in the horizontal plane but oriented in a particular direction (Schulson and Duval, 2010). As the ice freezes, pockets of brine form between the platelets. Most of this is initially within the grains, but it soon begins to drain in brine drainage channels. This drainage is from cold to warm, along the temperature gradient, generally downwards given a cold surface temperature. The brine content can be calculated from the temperature and salinity.

Multiyear (MY) ice is generally defined as having survived at least two summer seasons (but some sources define it as having survived a single melt season). It is typically 2 to 4 m thick and may thicken if more ice grows on its underside. Extreme ridge thicknesses can be many times the values just quoted and are a subject of probabilistic analysis for fixed offshore installations, in terms of the thickness, number of interactions and related parameters. The ridges will be smoothed compared to FY ice as noted above. We have discussed columnar ice structure; granular ice is composed of crystals with approximately equal dimensions. Brine inclusions may exist as irregular pockets at grain boundaries. Optical axes (or c-axes) are randomly oriented. Equiaxed crystals are crystals that have axes of approximately the same length, as may be the case with laboratory-grown ice. Equiaxed grains can often be an indication of recrystallization. "Statistical isotropy" is often a useful assumption in analysis. In this assumption, the material is treated as being an isotropic material, essentially by averaging over a set of crystals or grains. An analysis of S2 columnar and of granular ice with regard to the relation between the overall behaviour and that of single crystals has been given by Schapery (1997a).

Two other ice features should be noted as being of significant engineering importance as a potential hazard to ships and engineering. These are icebergs and ice islands. Icebergs are generally large masses of freshwater ice which have broken away from glaciers, a process referred to as "calving". Large icebergs are a potential hazard to offshore activities as are smaller pieces such as bergy bits especially if accelerated by waves in high sea states. A standard has been adopted by the IIP and the Canadian Ice Service for iceberg size classification. The corresponding waterline lengths are 0–5 m (growlers), 5–15 m (bergy bits), 15–60 m (small icebergs), 60–120 m (medium icebergs), 120–215 m (large icebergs) and greater than 215 m (very large icebergs). Icebergs can break up further, and, as noted, the resulting fragments also cause a potential hazard, especially in high sea states where they might be difficult to detect. Tabular bergs (flat-topped) and ice islands generally form from ice shelves. The ice islands may be up to 50 m thick and hundreds of square kilometres in area.

2 Ice Behaviour under Stress and Tools for Analysis

2.1 Reality of Ice Failure

As with other materials, testing of ice has often been carried out on small samples. This is as it should be, and the results are invaluable for our basic understanding. For some materials, for example those following a plasticity-based yield criterion, the failure surface can be modelled using the results of a single uniaxial test, usually in tension, to obtain the yield strength. In the case of ice, the situation is more complex. First, there is the question of time-dependency, so that more parameters are needed to model its response. Further complications arise, such as pressure-dependence, for which triaxial testing is needed. This is not modelled well by a pressure-dependent yield surface because of the time-independence of plasticity theory. Overriding these factors is the reality that small scale test results often bear little relation to failure stress, stress distribution, or failure mode found in field ice–structure interactions.

Reviews by Sanderson (1988) and Weeks (2010) show uniaxial compressive strengths of small scale specimens generally in the range 1–10 MPa. Small scale uniaxial tests have been reported by Sinha (1984) in an interesting programme of field testing that included first year and old ice. Strengths were in the range 1–5 MPa. Sanderson (1988) and Weeks (2010) also reviewed tests and analyses with respect to brine content, which tends to weaken the ice. They discuss tests by Wang (1979) on columnar grained ice with respect to the orientation of the c-axis, where loading parallel to the c-axis showed higher strengths than other orientations (note that the basal planes were then at right angles to the load). Tests on large ice specimens (with dimensions of 6 m × 3 m × 1.5 m thick) which were subjected to uniformly applied load at low rates showed no size effect, i.e. the strengths were similar to those found on small laboratory specimens (Chen and Lee, 1986; Wang and Poplin, 1986). These results are very useful for an understanding of ice behaviour and constituted excellent research but do not explain results obtained in the field. We know that ice failing against a vertical structure over a significant area (say of the order of 100 square meters), whether multiyear or first year ice, will impose average pressures much less than 1 MPa, and also impose high local pressures much higher than the highest uniaxial strength. There is no apparent agreement with the uniaxial strength nor any obvious way to link the two situations.

The answer to the questions just posed lies in the fact that the failure of ice acting in compression against a structure in a real field situation involves two highly significant localizations: first, of pressure into the localized high-pressure zones and second, the formation of a "layer" of microstructurally modified material adjacent to the structure within the high-pressure zones, not readily observed either in uniaxial testing or obtained in traditional continuum analysis. This situation is found particularly during ice crushing, a common failure mode observed in the field, even at ice speeds as low as 2.5 cm s^{-1} (Rogers et al., 1986). These zones of high pressure form and disappear in a birth-and-death process during ice crushing. The pressure distribution is far from uniform, with a parabolic shape across the high-pressure zone. This is in contrast to the uniform state of compression that would be found in a laboratory test on a cylindrical or prismatic specimen carefully machined to the desired shape. In the high-pressure zone, the peak stress is many times the uniaxial strength, with a complex triaxial state of stress, and with significant occurrence of microstructural change, especially microfracturing and dynamic recrystallization. The microstructural changes are highly time- and stress-dependent, as are the resulting stresses and strains, and they affect significantly the distribution of load and pressure on the structure applying the load. Figure 2.1 shows a representation of a typical localization, termed a "high-pressure zone" (*hpz*). These entities might be of the order of 30 cm across in field interactions of thick ice features, but are much smaller in thin ice sheets and small scale indentation. They occur at irregular intervals across the failure plane in ship– or structure–ice interaction, as illustrated in Figure 2.2, except at very low interaction speeds where creep processes predominate. Their formation is closely linked to fracture processes, which are not observed in uniformly loaded laboratory specimens. Fracture aspects still need considerable study to enable one to understand why the

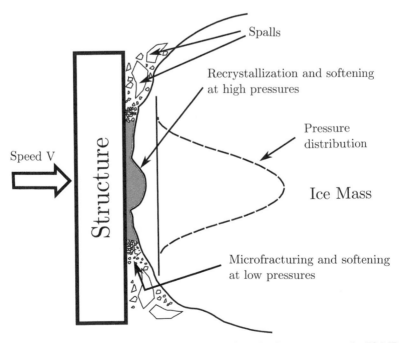

Figure 2.1 Schematic of a high-pressure zone (*hpz*). Peak pressures can be 70 MPa or higher.

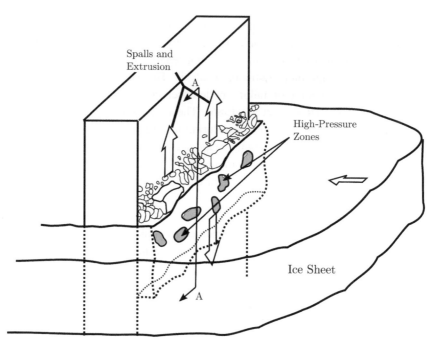

Figure 2.2 Schematic of formation of *hpz*s in interaction of structure with an ice sheet (Jordaan, 2001). With permission from Elsevier.

interaction area consisting of high-pressure zones constitutes only about 10% of the global area.

We shall analyse field data with regard to the forces imposed by *hpz*s in Part I of this work and discuss their formation and behaviour in some detail in Part II. Their existence should be borne in mind while reading the material to follow.

2.2   Creep, Fracture, Strength and Damage: Ice Crushing

Our main focus is the study of ice and its interaction with offshore structures and vessels. The emphasis is on thick ice features, multiyear ice or icebergs. Whether interacting with a ship or a structure, the ice will generally exhibit crushing failure. This rather unique failure mode was outlined in the preceding section and will be detailed throughout this work. It involves creep, fracture and changes to the ice microstructure. Ice deformation shows an extreme time-dependency. Such viscoelastic movements can relieve stress, yet ice is very brittle. The response of ice to stress depends strongly on the rate of loading and the stress level achieved. Creep is the time-dependent increase in the deformation of a material under stress. A concrete beam or slab will deform with time, occasionally resulting in unwanted deflections. Marble slabs have been known to deform considerably over time. Materials that creep include wood, many metals, steel at high temperatures, concrete, rocks, plastics, human tissues and ice. Materials at high values of homologous temperature are especially prone to creep, and ice is no exception. A specimen of ice loaded at constant stress will deform elastically and then over time deform by an amount that may be many times the elastic deformation. Under the compressive states of stress found in field observations of ice–structure interaction, the strain rate may be further increased by many orders of magnitude, which certainly occurs within *hpzs*.

The other dominant aspect of ice behaviour is fracture. The fracture toughness of ice is very small in value, one of the lowest values of materials on Earth. The formation of cracks will also involve time-dependent strains in the material, and as a result energy is consumed in viscoelastic deformation around the crack tip. This occurs in ice–structure interaction leading to time-dependent fracture, adding to the difficulty of analysis. A phenomenon related to this is "damage", in which the stress state and its history cause microstructural changes in the material, altering its mechanical behaviour. This area of research has been referred to as "damage mechanics", since in industrial applications, the effect on the performance of the material, e.g. steel, under stress, is of importance. In analysing a material such as ice, formed in nature, the word "damage" is rather inappropriate, but we retain it as the area of mechanics required is "damage theory" or "damage mechanics". The combination of time-dependency and brittleness makes ice a very interesting and challenging material to study, and may serve as a model for studies of other materials such as rocks and geophysical materials.

At low loading rates, ice under uniaxial compression deforms as a nonlinear viscoelastic material; at higher rates it experiences microcracking and microstructural change and at even higher rates fracture processes intervene. Uniaxial compressive strengths are typically in the range 1–10 MPa, depending on the loading rate, brine volume and other factors. Sanderson (1988) and Weeks (2010) have given good descriptions of the influencing factors. Tensile failure typically occurs at about 1 MPa. These, and other small-scale values obtained in the laboratory (or in the field), have, as noted in the preceding section, little relation to design pressures. A typical global pressure on a large structure (as measured say on the Molikpaq) might be considerably less than 1 MPa. Local pressures vary depending on the area, from say 70 MPa on areas the size of 1 cm^2 to perhaps 4–5 MPa on an area of 1 m^2. As noted in the preceding section, it is not possible to obtain such values from the measurements on uniaxial specimens; nor is it possible to deduce them from measured fracture toughness values.

During indentation and during ice–structure interaction, "crushing" failure occurs. As noted in Section 2.1, this is a special process involving fracture and consequent localization of pressure. In fact, the term "ice crushing" used to describe compressive failure of ice covers a multitude of phenomena: fractures occur in the form of flakes and spalls and many particles of crushed ice are ejected. A special process occurs under certain conditions, in which very small particles are ejected, often accompanied by larger flakes associated with spalling. The texture of the small particles is creamy and has been likened to toothpaste; such situations occur under certain conditions of interaction rate and temperature, and are associated with dynamic ice–structure interaction. The *hpzs* occuring during indentation have been found in medium scale tests conducted up to 3 m^2 in area but are also found in small scale indentation tests (with indentors generally 20 to 40 mm in diameter) making appropriate adjustments to the speed of indentation, as will be described in Section 8.6. It must be emphasized that the microstructural changes, when analysed in the context of the associated multiaxial state of stress and confinement, lead to profoundly altered mechanical response and behaviour, which cannot be ignored in the mechanics of ice response.

Strong analogies to the observations as summarized in the preceding paragraph can be found in the geophysical literature. Features that have a strong similarity to those just described are commonly observed. In rocks found near fault zones, cataclasis involves brittle fracture of mineral grains. This would correspond to the microfractured zones near the edges of the high-pressure zones in ice, noted in the preceding paragraph. Mylonites, on the other hand, involve reduction of grain size by a process of dynamic recrystallization, as in the centre of the high-pressure zones in ice. Sibson (1977) discusses these processes in detail. Many materials show time-dependent behaviour and also degrade with time. Concrete, ceramics, polymers and metals exhibit elasticity as well as time-dependent response to stress, and these materials may also exhibit microstructural change and microfracturing under stress, which often result in significant changes to the stress–strain behaviour, and in the case of ice, totally transform the response. All of the behaviours discussed—creep, fracture and damage—are present in ice and geophysical materials under stress. In the case of ice, these phenomena occur in a rather complex yet interesting combination.

2.3   Approach to Mechanics

Given the need to deal with both elastic and time-dependent (dissipative) deformations, it is important to develop approaches that are soundly based on thermodynamic principles. Mechanics is based on the concepts of mass, length and time, from which we derive force and displacement, stress and strain, work and energy, and related quantities. Temperature and heat are added in thermodynamics. The intuitive ideas of a spring and a dashpot are used as a basis, with material degradation over time leading to a change in material properties and to the response of the spring and dashpot to stress. The work of Biot and Schapery as introduced in Part II and Appendix A provides the foundation to the approach, based on the thermodynamics of irreversible processes (TIP). Bridgman (1941) gives very useful perspective on the first law from the operational point of view that he advocates. If one is dealing with a body, and we are considering a bounding surface around the region of interest, one imagines "sentries" posted at points around the boundary, each with instruments capable of measuring the amount of heat passing through their element of the bounding surface, or the flux of mechanical energy by means of measurement of stress and displacements. Much of this can now be studied using numerical and finite element analysis. The flux of mechanical energy into a small volume of material is causative in the microstructural changes to the material, which effectively alter its mechanical properties, especially its time-dependent response.

In large ice masses, such as multiyear ice or icebergs, the ice crystals are found in an irregular array of more or less random arrangements, but fitting together. The process to be analysed leads to breakdown of the original structure by microfracturing and by dynamic recrystallization or by fracture. In processes involving dynamic recrystallization, the ice crystal structure develops a granular, equiaxed morphology. The crystals will initially be of different sizes, and any attempt to follow the fate of individual crystals would quickly lead to an intractable problem, given the objective of studying the full failure process of the ice feature up to and after the occurrence of peak load. Much of the analysis of mechanics in Part II has been conducted using the finite element program ABAQUS and associated user material subroutines UMAT and VUMAT. Ice is in general treated as being statistically isotropic. The condition (state) of particular points in the continuum have been represented by internal state variables. With regard to fracture, the approaches of Barenblatt as extended by Schapery to include time-dependence are the most appropriate for the present work.

A statement of the first law of thermodynamics has been given by Bridgman (1941):

$$\Delta u = \Delta h + \Delta w, \tag{2.1}$$

where Δu = gain of internal energy for the region of interest, Δh = heat received and Δw = work received by the region from its surroundings, during any time interval. For the purposes of the analyses in the present work, entropy S is associated with the generation of heat h in processes including viscoelasticity, sliding of particles, frictional movements and dynamic recrystallization. Grain refinement can lead to substantially increased viscous flow. Entropy change dS is given by

$$dS = \frac{dh}{T}, \tag{2.2}$$

where T is temperature. The unit of entropy is J K^{-1}. There are small differences to be recognized in the elastic properties under isothermal versus adiabatic conditions; these are summarized in Appendix A.

2.4   Creep of Ice

In introducing the creep of ice, we first make a connection to the work of Andrade (1910). He reported on experiments of tensile creep of lead, lead-tin alloys, and copper, and observed that the deformation could be divided into three parts: (i) the immediate extension upon loading; (ii) an initial flow which gradually disappears; and (iii) a constant flow, taking place throughout the extension. It is most useful to distinguish between elastic strain, delayed elastic strain, and flow, as these can be related directly to viscoelastic theory. The strain due to a constant small stress that is applied uniaxially and removed after a period of time is shown in Figure 2.3. We write the strain under constant uniaxial stress as

$$\epsilon = \epsilon_e + \epsilon_d + \epsilon_f, \tag{2.3}$$

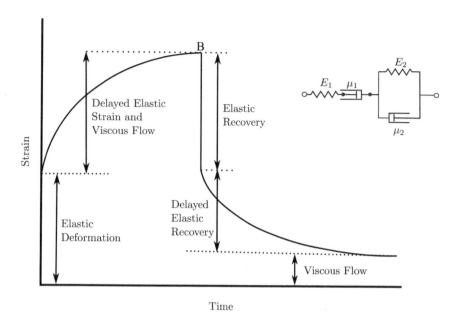

Figure 2.3 Components of creep for low stress. Stress applied at the origin of the time axis and removed at time corresponding to point B shown in figure. The Burgers unit is illustrated.

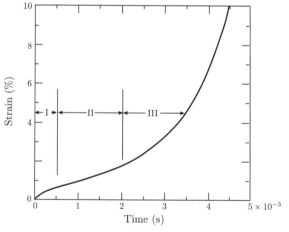

Figure 2.4 The three stages of creep at higher stress levels: primary, secondary and tertiary (after Mellor and Cole, 1982, with permission from Elsevier).

where ϵ_e, ϵ_d and ϵ_f are the elastic, delayed elastic and the flow components of strain. The elastic strain is recovered immediately upon unloading, whereas the delayed elastic component recovers over a period of time. For higher stresses and longer periods of time, there are three approximate stages: primary (I), secondary (II) and tertiary (III), as shown in Figure 2.4. In the last of these stages, the strains accelerate, largely associated with damage processes. Ice does exhibit the minimum creep rate as shown in Figure 2.4, given a high enough level of stress. Conceptually, the deformation of Figure 2.3 can be modelled as a Burgers unit (Figures 2.3 and 6.7)—more on such models in Chapter 6. The third component of this subdivision ϵ_f we term "flow", irrecoverable viscous strain. In ice there will inevitably be microstructural change except for deformations at very low stresses and for short durations of loading. Elasticity involves largely reversible processes, while the flow term, and dashpots in general, involves dissipation and increases in entropy. The elasticity of the ice crystal is detailed in several references, including the recent work by Schulson and Duval (2009), who also discuss creep deformation of single crystals as well as polycrystalline ice.

We shall adopt the equations of Sinha (1978, 1979), which follow this subdivision of strain. We deal with compressive states so that we take these as being positive, with associated contractive deformations as being positive also. Sinha's work was concerned with low stresses, generally less than 1 MPa, and with uniaxial loading of the specimens. Various temperatures were considered including a base for comparison of response at $-10°$C. His equation for strain under constant stress reflects the three components of strain as discussed:

$$ \epsilon(t) = \frac{\sigma}{E} \left\{ 1 + \frac{c_1 d_1}{d} \left[1 - \exp\left(a_T t\right)^b \right] \right\} + \dot{\epsilon}_0 t \left(\frac{\sigma}{\sigma_1} \right)^n, \qquad (2.4) $$

where σ is the stress, E is the elastic modulus, d is the average grain diameter and c_1, d_1, a_T, b are constants. The term for flow, derived essentially from Glen (1955), contains the constants $\dot{\epsilon}_0, \sigma_1$ and n; Glen found $n \simeq 3$. The latter relationship is Glen's law. For a discussion of equations for creep of ice, see also Ashby and Duval (1985).

The mechanisms involved are associated with movements of dislocations through the crystal lattice, and Sinha postulates grain boundary sliding as the mechanism associated with delayed elastic strain. It is noted in Duval et al. (1983) that non-basal slip requires a stress of at least 60 times that of basal slip at the same strain rate. This anisotropy and resulting distribution of internal stresses were analysed, based largely on experimental observations. The redistribution was interpreted by Duval et al. (1983) as the basis for delayed elastic strain. The relationship between single-crystal behaviour and polycrystalline ice was studied later by Schapery (1993, 1997a) with a focus mainly on S2 ice. The elastic properties were estimated using self-consistent and bounding methods, and the results were extended to viscoelastic analysis using the elastic-viscoelastic analogy (see Section 6.9). The use of viscoelasticity is appropriate and the reaction between the crystals was analysed for basal-plane shearing. This was carried out with a power-law input. The use of activation energy is outlined in Section 2.8.

We can write equation (2.4) as

$$ \epsilon(t) = \frac{\sigma}{E} + \frac{\sigma}{E_k} \left[1 - \exp\left(a_T t\right)^b \right] + \frac{\sigma}{\mu(\sigma)}, \qquad (2.5) $$

where E_k is the modulus for delayed elastic strain, and $\mu(\sigma)$ is a stress-dependent viscosity $\propto 1/\sigma^{n-1}$. The two terms describing creep in equation (2.5) are the primary and secondary creep components. It is important to distinguish the linear terms in equations (2.4) and (2.5), as we wish to know whether to use methods of linear viscoelasticity. The elastic modulus E and the coefficient $E_k = (c_1 d_1/d)$ indicate linear elastic and delayed elastic terms, yet a complication is presented by the exponent b

in the equation for the latter. A linear viscoelastic equation for delayed elastic strain would typically correspond to the form $b = 1$. See Chapter 6. Note that for small t,

$$1 - \exp{(a_T t)^b} \simeq (a_T t)^b. \tag{2.6}$$

This is similar to Andrade's expression for the "flow which gradually disappears", which is written as

$$\beta t^b, \tag{2.7}$$

where t is time and β and b are constants, with $b \simeq 1/3$. He termed this β-flow and acknowledged that it should have a finite limiting value at large times. The expression βt^b does not have such a limit but provides excellent fits over decades of values of strain. It is a linear viscoelastic term with a distinct spectrum (see Section 6.8).

We can therefore conclude that, at least for small time intervals during primary creep, ice may be treated as a linear viscoelastic material. This approximation is valid for about 100 seconds, during which time secondary creep is small or negligible. The primary creep eventually reaches a limit. The rate after this stage corresponds to the "minimum creep rate" of secondary creep, as shown in Figure 2.4. The creep rate then accelerates. Ashby and Duval (1985) analysed a variety of data sets and developed a model for polycrystalline ice that agreed well with data up to the minimum creep rate. Schapery (1991a) re-analysed the data using a model based on the "modified superposition principle" (MSP) which included the effects of damage (microstructural change) and showed that good agreement with data could be achieved, including the accelerating creep rate in tertiary creep. Indeed, it was suggested that the minimum creep rate itself arises from damage processes.

The creep relationship of Sinha as introduced above provides an adequate basis for our discussions, especially as the strain is divided into the three components as observed. This is most useful for our introduction of viscoelasticity in Chapter 6. We emphasize that the states of stress in actual ice–structure events in the field are a far cry from the uniaxial low-stress conditions employed in creep testing. These conditions simply do not extrapolate at all to the actual conditions when ice interacts with a structure.

2.5   Pressure Melting of Ice

The fact that water expands upon freezing leads to the phenomenon of pressure melting: when pressure is applied to ice, the melting temperature reduces. For example, a pressure of about 135 MPa would decrease the melting point to $-10°C$. One of the earliest applications of thermodynamics was concerned with this subject, going back to Clapeyron and James Thomson. Experimental data on this question were presented by Bridgman (1912). The phenomenon has profound consequences with regard to the behaviour of ice in compression. We first note that when a piece of ice melts, its volume is reduced. As a result, work is done on it by atmospheric pressure. The ice requires the provision of latent heat equal to L per unit mass to melt, and part of this is provided by the work just noted, i.e. the atmospheric pressure times the change in volume. If the piece of ice at a given temperature, T_0, solid but close to melting, subjected to a lower temperature, it remains solid. But if an additional pressure is applied, for instance by mechanical means, then more work is done by the pressure during melting and less heat needs to be supplied to make up the latent heat. Melting then takes place at a temperature that is lower than the original given temperature T_0. As a result, melting will take place if sufficient pressure is applied; without the additional pressure, melting would not take place.

More formally, we consider a small amount of ice, say 1 g, under hydrostatic pressure p and with latent heat of fusion L. The ice is at its melting point T. As it melts, it contracts by the amount $(v_i - v_w)$, where v is the specific volume and the subscripts i and w refer to ice and water, respectively. The work done on the ice as it melts is

$$p(v_i - v_w). \tag{2.8}$$

If a small additional pressure is applied, increasing to the value $p + dp$, there is an additional amount of work on the ice during melting, equal to

$$dp(v_i - v_w). \tag{2.9}$$

The entropy change for melting is

$$S = \frac{L}{T}. \tag{2.10}$$

If the melting temperature changes by dT, we can estimate the entropy change from

$$\frac{dS}{dT} = -\frac{L}{T^2}, \text{ and} \tag{2.11}$$

$$T\,dS = -\frac{L}{T}dT. \tag{2.12}$$

Equating this heat to the additional work,

$$dp(v_i - v_w) = -\frac{L}{T}dT, \text{ and} \tag{2.13}$$

$$\frac{dT}{dp} = -\frac{T(v_i - v_w)}{L}. \tag{2.14}$$

This can be derived also by considering the Gibbs free energy

$$G = u + pv - TS, \tag{2.15}$$

where u is the internal energy and S, as before, the entropy. At equilibrium, the specific Gibbs free energies of the solid and liquid are equal. When we add heat at constant volume, this becomes

$$-S_i\,dT + v_i\,dp = -S_w\,dT + v_w\,dp, \tag{2.16}$$

and, as a result

$$\frac{dT}{dp} = \frac{v_i - v_w}{S_w - S_i}, \text{or} \tag{2.17}$$

$$\frac{dT}{dp} = -\frac{T(v_i - v_w)}{L}, \tag{2.18}$$

as before. Equation (2.18) is the Clausius-Clapeyron relationship. Using equation (2.18), the accepted result for ice at 0°C is a decrease of 0.0738°C MPa^{-1} (Hobbs, 1974).

Bo Nordell (1990) designed an ice-powered cart, the driving force of which was obtained from the pressure-volume work of freezing water. This developed into an interest in pressure melting. He carried out experiments and extended equation (2.18) to take into account the thermal expansion and volume change at constant pressure and constant temperature. He obtained a nonlinear relationship replacing equation (2.18), which agreed well with his experimental results and those of Bridgman and co-workers. The linear relationship of equation (2.18) with the value of 0.0738°C MPa^{-1} gives good results up to 50 MPa, and deviating (underestimating the decrease in temperature) by about 7% at 100 MPa. Although surface pressures up to about 70 MPa have been measured in indentation tests, these values would be the sum of volumetric and deviatoric components. The associated hydrostatic pressures would be of the order of 40 MPa. Nordell's test results show that a confining pressure of about 110 MPa is required to lower the melting temperature to −10°C. The International Association for the Properties of Water and Steam gives refined relationships. It is interesting to note that, with increasing pressure, the melting point decreases to a minimum of about −22°C at 210 MPa. Thereafter, the pressure melting point rises with pressure, passing back through 0°C at 632 MPa.

2.6   Dynamic Recrystallization

Dynamic recrystallization refers to the nucleation and growth of new grains of a material during deformation. The adjective "static" is reserved for recrystallization that is not associated with the deformation. Dynamic recrystallization (DRX) is a softening mechanism in metals and alloys during deformation (under stress) at elevated temperatures. Strong decreases in grain size may lead to diffusive mass transfer (Urai et al., 1986). The area of interest in the present work is the dynamic recrystallization of ice under stress, whereby a strong decrease in grain size does indeed result. In the high-pressure zones (*hpzs*) referred to earlier in Section 2.1, high values of shear stress combine with a range of hydrostatic pressures, from low to high. Results regarding this phenomenon are discussed in many sections of this work, particularly in Section 4.7 on medium scale testing and in Chapter 7 on triaxial testing. The process leads to extensive microfracturing combined with recrystallization at low hydrostatic pressures and extreme grain refinement at high hydrostatic pressures. Thin sections of ice illustrating this process are given in Chapter 7; reference is also made to Melanson et al.

(1999), Meglis et al. (1999) and Jordaan and Barrette (2014). At the centre of the *hpz* during deformation under high shear and pressure, the dominant deformation mechanism is much enhanced flow, certainly nonlinear with stress. An extreme softening process takes place in ice; the effects of pressure melting add to this.

The changes in structure are driven by mechanical processes and the use of free energy (e.g. Cottrell, 1975). This will be principally elastic strain energy; in Jordaan et al. (2005b), the strain energy and dislocation density with associated energy were estimated. Dislocation density is described as leading to dynamic recrystallization in metals, when the metal is raised to the recrystallization temperature. Alaneme and Okotete (2019) reviewed the mechanisms of recrystallization and noted lattice strains and crystalline imperfections as being important, with dislocations being the main contributors. Eshelby (1971) states that the "effective force on crack tip" = J integral = "energy release rate is closely related to the force on a defect (dislocation, impurity atom, lattice vacancy and so forth)". The states of stress vary considerably, being under triaxial compression in a high-pressure zone. No doubt, these states of stress, when superimposed on material containing dislocations, will increase the local stress concentration and energy. Stress concentrations between grains will lead to localized pressure melting. The fine-grained structure and associated strong decrease in grain size lead to a change in deformation mechanism from creep to a process where finely grained particles are extruded, with extremely small pieces of ice moving and mixing with others. The mechanical behaviour of the microstructurally modified ice has been measured in a series of experiments using a triaxial cell (Section 7.4).

With regard to the geophysical analogue of this process, mylonite formation is discussed in Urai et al. (1986). Etheridge and Wilkie (1979) suggested that dynamic recrystallization alone is insufficient to cause enough weakening to result in flow of mylonites. This is certainly not the case for ice, in which deformation is also aided by pressure melting. Whether this is a factor in any mineral deformation is a subject for consideration. Very high pressures and temperatures do exist in mylonite formation; Garlick and Gromet (2004) state that diffusion creep does occur with partial melting in the formation of high temperature mylonitic gneisses. These few words are not intended to be a survey of literature; the purpose is to point out the analogy between the processes in ice under compressive stress and those in mylonite formation (at vastly different scales).

2.7   Microstructurally Modified Ice: Development of Layers

The "layer" of ice, a seemingly minor detail of compressive ice–structure interaction, turns out to be highly significant in understanding compressive ice failure. Under compression, especially those non-uniform states that occur in interaction with structures or under indentation, ice is prone to develop zones with microstructurally modified texture, and particularly zones of recrystallized ice. We introduce these generally in this section (see also Jordaan et al., 2009) with regard to past experiments, and return to the subject in discussing medium scale indentation in more detail (Section 4.6).

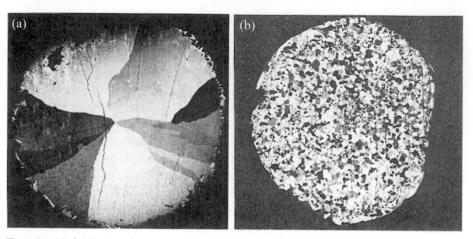

Figure 2.5 (a) Original grain structure of ice specimens of bubble free ice. (b) microstructure of bubble-free ice extruded at a strain rate of 5.6×10^{-2} s^{-1} at a temperature of $-20.5°$C. From Kuon and Jonas (1973). This content has been reproduced with the permission of the Royal Society of Canada.

Barnes et al. (1971) studied the friction on a cone of single-crystal ice sliding over a hard surface. They observed recrystallization to occur at the interface. The formation of this zone of recrystallized ice helped account for the increased rate of creep found in the friction experiments versus comparable compression experiments. Offenbacher et al. (1973) studied recrystallization of single crystals of ice using thin ice specimens and a transparent glass surface, which could be moved at speeds ranging from 0.05 to 4 mm s^{-1}. By observing the contact zone between the ice and glass plate, it was noted that at loads above about 1.5 kg (or at lower loads that were held for longer periods of time), "permanent changes occurred immediately", with diameter of the recrystallized area about 2 mm. Cracks were often produced in a direction perpendicular to the c-axis. Recrystallization was found to begin along these cracks and then spread into the deformed region of the ice. The inadvertent presence of a grain boundary caused similar recrystallization behaviour to occur.

Kuon and Jonas (1973) performed tests aimed at studying the effects of strain rate and temperature on the extrusion behaviour of ice. These authors tested samples of ice with and without air bubbles. The test specimens were 4.12 cm in diameter and 20 cm in length. The bubble-free specimens consisted of large columnar grains extending from the outside to the centre of the sample. In the case of the ice containing bubbles, the ice had a columnar structure over most of the sample with a central core containing equiaxed grains. Tests were performed by extruding the specimens of ice through a die and hollow ram. It was found that at constant extrusion temperature, the grain size decreased with increasing strain rate. At constant strain rate on the other hand, the grain size increased with temperature. In all cases, considerable reduction in grain size occurred; Figure 2.5 shows one result. The authors speculated that subgrains observed within some of the recrystallized grains were formed during the deformation and consequently were caused by dynamic recrystallization. Frederking and Gold

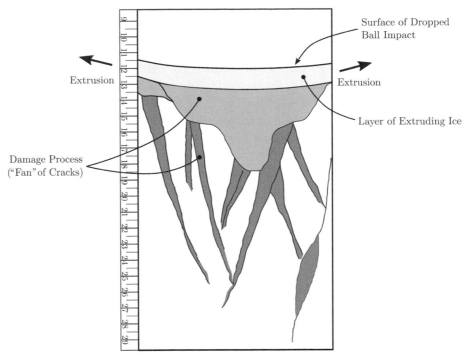

Figure 2.6 Writer's representation of ice section after dropped-ball test by Kheisin and co-workers. See also original paper by Kheisin and Cherepanov (1970).

(1975) observed a reduction in grain size in regions of high shear during laboratory experiments on indentation of ice plates within the range of ductile behaviour, i.e. at slow rates of interaction.

At high rates of indentation or interaction, a distinct layer of microstructurally modified ice is found. Kheisin and his co-workers (Kheisin and Cherepanov, 1970; Kurdyumov and Kheisin, 1976) described tests conducted in 1967 using a steel ball weighing 300 kg dropped onto the ice surface (freshwater ice of Lake Ladoga). They examined the microstructure before and after the impact. Velocities were in the range 1 to 6 m s^{-1}. They drew attention to the fine-grained nature of the ice in a layer, which covered the entire contact zone. The main features of the results are illustrated in Figure 2.6. The distinct layer is the feature that was exploited in further analysis, and they also noted "fans" of cracks further into the ice. Kheisin and co-workers described the changes to the microstructure in some detail and pointed out that "lamellae up to 0.5 mm thick, formed by shear over the basal planes" were produced. They also pointed to the existence of "ice particles of submicroscopic size dispersed amidst these crystals and acting as a lubricant for the matrix", resulting in a "finely dispersed crushed layer behaving as a layer of viscous liquid". The effect of pressure melting and frictional heating was also described.

In the present work, considerable attention is paid to medium and small scale testing, in which the state of the ice could be examined after testing. Layers of microstructurally modified ice were invariably found, except at very low indentation speeds. We shall return to discussion of the medium scale tests in Section 4.7 and the small scale tests in Section 8.6. A most important fact that emerges is that the behaviour of ice in these crushed layer is quite different from that of the parent ice. Extrusion of ice particles on a significant scale is a feature of this failure mode, an aspect that is hard to reconcile with conventional mechanics. Time-dependent effects and changes to structure requiring viscoelasticity and damage mechanics are dominant. In the case of the layer, as studied by Kheisin and his colleagues, it was treated as being viscous, distinct in properties from the parent ice. This acknowledgement of reality was a big step forward in ice mechanics, as the distinction between the properties of ice in an unmodified state, on the one hand, and ice that is microstructurally modified and under high compression, on the other hand, is really important to the understanding of how ice fails in compression. The approach taken by Kheisin and co-workers was to treat the layer as a linearly viscous material with its own viscosity. This approach is described in outline form in Section 6.4.

2.8   Thermally Activated Processes

Many processes can be modelled as being thermally activated: movements of dislocations in metals and other materials, structural changes including recrystallization and fracture processes are examples (Krausz and Eyring, 1975). We can certainly conceive of creep movements as being thermally activated. The material is in equilibrium which is disturbed by repeated thermally activated events. A particle that acquires an energy equal to or greater than the activation energy e_A is able to surmount an energy barrier and move into a position of lower potential energy (Cottrell, 1964, 1975). These movements are observed macroscopically as creep of the specimen.

We consider a particle oscillating about a certain mean position. Let E be the event of a particle surmounting the energy barrier e_A in a "run" at the barrier. Noting that the energy level $e_i = ih\nu$, where h = Planck's constant, ν = frequency, and $i = 1, 2, 3,....$, the classical Maxwell-Boltzmann distribution of energy amongst particles (summarized in Jordaan, 2005) is expressed as follows:

$$p_i = \frac{\exp\left(-e_i/kT\right)}{\sum_j \exp\left(-e_j/kT\right)}. \tag{2.19}$$

Then the probability of event E is given by

$$\Pr\left(E\right) = \frac{\sum_{e_i>e_A} \exp\left(-e_i/kT\right)}{\sum_j \exp\left(-e_j/kT\right)} \tag{2.20}$$

$$= \frac{\sum_i \exp\left[-\left(e_A + ih\nu\right)/kT\right]}{\sum_j \exp\left(-jh\nu/kT\right)} \tag{2.21}$$

$$= \frac{\exp - \left(e_A/kT\right) \sum_i \exp\left[-\left(ih\nu\right)/kT\right]}{\sum_j \exp\left(-jh\nu/kT\right)} \tag{2.22}$$

$$= \exp\left(-\frac{\epsilon_A}{kT}\right). \tag{2.23}$$

If v = the number of runs per second at the barrier, the rate of a spontaneous reaction using the theory of activation energy is

$$v \exp\left(-\frac{\epsilon_A}{kT}\right). \tag{2.24}$$

This may be written as a frequency f per second

$$f = v \exp\left(-\frac{Q}{RT}\right), \tag{2.25}$$

where Q is the activation energy per mole of particles and R is the gas constant. This may also written as strain rate:

$$\dot{\epsilon} = A \exp\left(-\frac{Q}{RT}\right), \tag{2.26}$$

see for example Glen (1955). The term A is called the "pre-exponential factor".

This relationship has been used extensively in the literature on creep of ice, for example Barnes et al. (1971), Glen (1955) and Sinha (1978). One can plot log $\dot{\epsilon}$ against the reciprocal of absolute temperature to test the relationship and to use it to making inferences regarding the value of Q. This was done by Barrette and Jordaan (2003) for the minimum creep rate under various triaxial loadings. This is discussed in Section 7.5. Schapery (1996, 1997b) included discussion of molecular models in nonlinear viscoelastic analysis.

2.9 Fracture Processes

Fractures are omnipresent in ice mechanics. Gold (1963) recognized the importance of the extreme brittleness of ice. Palmer et al. (1983) correctly emphasized the role of fracture in estimating ice forces. There have been several good test programmes of measurement of fracture toughness. Yet there has been little progress in development of analytical tools based directly on fracture mechanics for use in engineering design. To take the most prominent example of the effect of fracture, we consider compressive "crushing" failure. We know from measurements that the load transmitted to the structure is concentrated in a small percentage, of the order of 10%, of the area defined by the global area, obtained from the original unfractured profile of the ice. The load becomes localized into many high-pressure zones (*hpz*s), as illustrated in Figures 2.1 and 2.2. Design methods are generally based on an empirical analysis of measured pressures. A method that is defensible on a probabilistic basis is one in which *hpz*s are modelled as a random set of loads at spacing that is partly random (e.g. spaced randomly but near the centre of ice sheets). This is described in Section 3.4 for the case of ships ramming thick ice. Fracture mechanics has not fully explained the very important spalling events that cause the localization and the resulting 10% of global area that constitutes the loaded area.

A further complication is that, as with other aspects of ice deformation, fracture is time-dependent, with viscoelastic movements highly involved in the fracture process. Strong evidence in the field was found during the medium scale testing that slow indentation results in a permanent "dent" in the ice with little or no spalling, whereas for faster loading a considerable amount of spalling occurred with *hpz* formation. Yet slower loading of longer duration can lead to sudden large fractures; this is described in Section 4.7. Icebreaker captains have reported that "leaning on the ice" can lead to time-dependent fracture of an ice floe. Large floes have been observed to split into pieces after a period under load (Figure 4.8). Reference is made to Chapter 9 for an analysis of time-dependency. As a result it is difficult to use the notion of "ductile to brittle transition" which has been borrowed from study of the behaviour of metals with regard to temperature. In ice mechanics "transition" is used with regard to strain or loading rate. In reality ice behaviour is more of a continuum. There is a transition to layer formation and extrusion from more widespread damage formation. Even in uniaxial testing, macroscopically "ductile" behaviour can result from a distribution of microcracks.

The theoretical strength of a material can be calculated as the stress required to break atomic bonds, and calculations show this to be of the order of $E/10$, where E is the elastic modulus. In the case of ice, this would give values of the order of 900 MPa, much greater than the tensile strength of ice, which is of the order of 1 MPa. As is well known, the difference is related to flaws, irregularities and defects in the material. The flaw that is studied in fracture mechanics is generally the crack. Depending on the state of stress in the material, the crack can be subjected to tensile stress, resulting in the "opening mode" of crack separation, or to shearing stresses, resulting in the "sliding" and "tearing" modes. We shall focus on the opening mode in the present work. To motivate an approach to fracture, we can think conceptually of a body with a crack of length $2a$, a being the half-length, with a tensile stress applied. Then the failure stress σ_f is related to crack length by the following:

$$\sigma_f \propto K_c / \sqrt{a}, \tag{2.27}$$

where K_c is the fracture toughness, a material constant. The longer the crack, the smaller the failure stress, which accords with common experience. Approximate values of fracture toughness of various materials (in kPa $\sqrt{m}$) are as follows: steel, 50,000; stainless steel, 200,000; aluminium, 20,000; ceramic, 4,000; soda lime glass, 750; and PMMA, 1,000. Timco and Frederking (1986) performed early measurements on ice in which fracture toughness values of the order of 100–140 kPa $\sqrt{m}$ were reported, very small values compared to other materials. The resulting energy release rate, taking the elastic modulus as 10 GPa, is in the range 1–2 J m^{-2}. Results on fracture toughness are reviewed in more detail in Chapter 9.

Events such as floe splitting (Figure 4.8) have received some attention with regard to load estimation but it is difficult to include these events in design calculations given present knowledge. It is important to study such events in future work using fracture mechanics, especially fractures induced in icebreaking. In the case of a large ice feature, such as a multiyear ice floe interacting with a structure, the large fractures

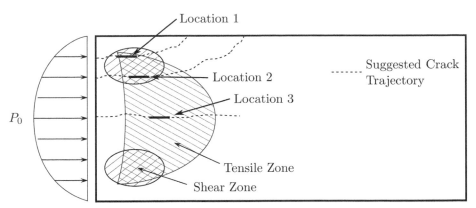

Figure 2.7 Possible fracture and spall zones shown in vertical section of beam under load (from Zou et al., 1996, with permission from Elsevier).

might relieve load (but likely not immediately); to put this into a probabilistic design context with appropriate mechanics, one has to know what is the probability of the fracture happening, where it might happen, and what stress relief results. The stress relief might be quite small in the case of a wide structure. While the fracture of large ice features is certainly of interest, other fracture events deserve attention, especially those involved in the formation of high-pressure zones.

It seems clear that the fractures associated with the formation of *hpz*s are relatively small—not of the size exhibited in floe splitting, or in fracture of large floes in icebreaking—but rather of the order of 1 or 2 metres, more or less. With regard to the formation of spall fractures and consequent localization, some studies have been initiated. Matskevitch and Jordaan (1996) related the confinement and state of stress that occurs in *hpz*s to the outwardly sloping sides observed in *hpz* formation. This has been illustrated in a general way in Figure 2.1. The sources of crack initiation in a stressed beam were explored by Zou et al. (1996), and the results are illustrated in Figure 2.7. Dempsey et al. (2001) considered the *hpz* as a hollow cylinder with pressure applied internally and externally. Pressure within the hollow cylinder is applied to the material, and the authors studied crack propagation from the inner cylinder, a very idealized situation. Taylor and Jordaan (2015) conducted a detailed investigation, the basis of which is illustrated in Figure 2.8. A spatial density of cracks was considered, based on grain size (Cole, 1986), with a random orientation. A simulation procedure was followed, whereby cracks were modelled and investigated as to their propensity to propagate. Figure 2.8 shows the initiating flaw and the stress field. This was carefully studied for various locations of crack appearance. Using LEFM, a wide range of situations were modelled: sheet ice of various thickness, tapering edges and eccentricity of loading. The results are promising and should form a good basis for further research. It is noteworthy that a scale effect with regard to thickness was found, following (thickness)$^{-0.50}$.

The localization into *hpz*s is in engineering practice handled empirically by using measured data to estimate local and global design pressures, as described in Chapters 3–5. Treating *hpz*s as random localized loads with random spacing can capture well

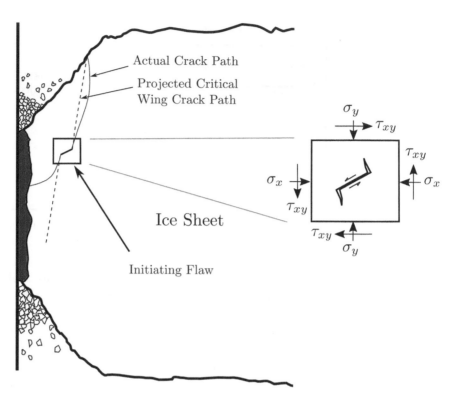

Figure 2.8 Schematic of analysis by Taylor and Jordaan (2015), with permission from Elsevier.

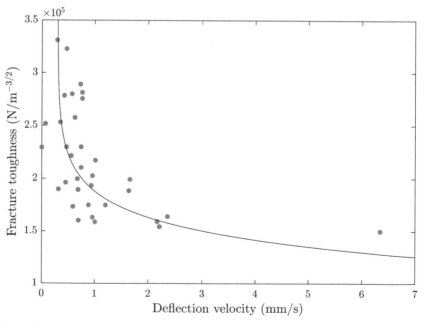

Figure 2.9 Variation of apparent fracture toughness with rate of displacement in four-point bending test on ice samples (Kavanagh et al., 2015) with power-law fit. With permission from ASME.

the observed behaviour and distribution of pressure for design purposes. The spacing should reflect the geometry of the interaction; for thin sheets for example, the loading is oriented along a line near the centre, the well-known line-like loading (Section 4.2).

Fracture is discussed in more detail with emphasis on time-dependent effects in Chapter 9. To illustrate the importance of time, Figure 2.9 from Kavanagh et al. (2015) shows an almost 3-fold variation of apparent fracture toughness. Care must be taken to understand time-dependent effects in ice fracture generally and in particular in postulating a scale effect of fracture toughness. The early work on cracks concentrated attention on the elastic stresses at crack tips, which were analysed as infinite, associated with a singularity. This was overcome partly by Griffith who considered the energy transferred in crack extension rather than crack tip stresses, and by Barenblatt, who eliminated the crack-tip singularity, using linear elastic theory. This was taken up by Schapery to study time-dependent effects and nonlinearity. These matters are considered to be research topics, and are deferred to Chapter 9 in Part II of this book.

3   Design Basics and Ship Trials

3.1   Local and Global Pressures: Exposure

There is nothing more useful in design than past measurements at a scale as far as possible approaching that of the structure or installation under consideration. For this reason, a comprehensive review of available full and medium scale measurements is given in this chapter and Chapter 4. In particular projects, access to original data is most useful, if this can be obtained. The full scale measurements provide vital information on localization of force into *hpz*s and associated values of force and pressures. A discussion of the history of pressure analysis for design has been given in Masterson et al. (2007). Initially, small scale data were used. During the 1980s, field tests were conducted at a larger scale. Data were also obtained from measurements on the hulls of structures, specifically the Molikpaq and on ship hulls during dedicated ramming tests. These data strongly indicated a size or area effect, in which the pressure decreased as the loaded area increased. The relationship is usually expressed by the power-law expression:

$$p = ka^{-n}, \tag{3.1}$$

where p is the ice pressure (MPa), a is the loaded area in square metres and k and n are constants, with n being less than 1. A strong lesson was that full scale measurements are much more useful than small scale ones in engineering practice.

As the methods for design developed, it was found necessary to define two areas: one for global design and the other for local design. Global design refers to consideration of forces applied to the structure as a whole, for instance in determining the resistance of a gravity-based structure to sliding or overturning, or in considering the maximum hull girder force in ship ramming of heavy ice. Local design refers to consideration of parts of the structure, for instance the design of plates between frames against excessive deflection or rupture. These two aspects of design require consideration of two definitions of area (Jordaan et al., 2005a). The *global interaction area* (which has in the literature also been termed the *nominal interaction area*) is the area determined by the projection of the structure onto the original shape of the ice feature, without any reduction of the area for spalls and fractures that take place during the interaction. Figure 3.1 shows a schematic of the plane of interaction between an iceberg and a structure. The global interaction area can be determined from the projection of the shape of the iceberg or ice feature onto the structure form. Within this area, there

(a)

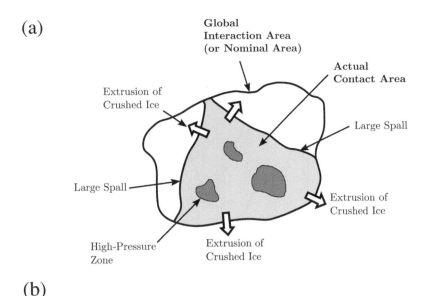

(b)

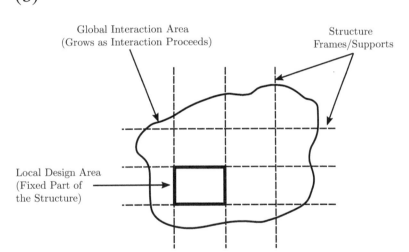

Figure 3.1 Schematic illustration of an interaction plane between an iceberg and a planar structure. (a) Global area. (b) Local area for design.

will be areas that carry little or no pressure, as well as high-pressure regions. When analysing data to formulate relationships for global load estimation, the analysis must also be done in terms of the proper global interaction area.

There are smaller areas (e.g. containing *hpz*s) within the global interaction area that are subjected to high local pressures, considerably higher than the global average pressure. For design purposes, one needs to consider the *local design area*, which is the area of the part of the structure to be designed, for example a panel or the plating between frames. A separate model has been developed to estimate pressures on a local design area. Since the area for local pressure, defined in the lower part of Figure 3.1,

(a)

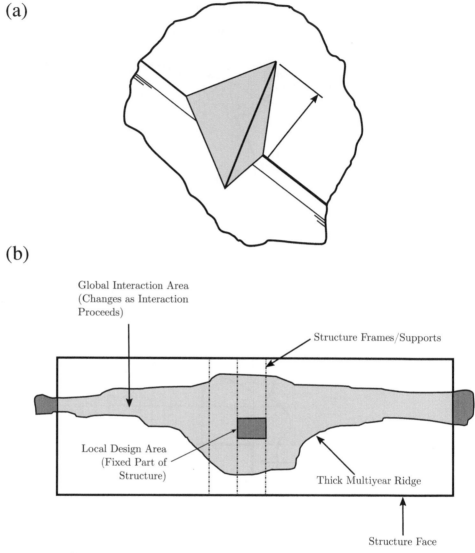

(b)

Global Interaction Area
(Changes as Interaction
Proceeds)

Structure Frames/Supports

Local Design Area
(Fixed Part of
Structure)

Thick Multiyear Ridge

Structure Face

Figure 3.2 Global area for ship penetrating an ice floe (a); local and global areas for ice ridge interaction with a fixed structure (b).

is that of a fixed area in the structure, an ideal measurement for use in design is that corresponding to an instrumented panel in a ship or structure.

The appropriate area to use in the determination of global average ice pressures in ship–ice interaction is shown in Figure 3.2 (a). See also Riska (1987). Frederking (1999) reviewed pressure–area relations and defined the *process* pressure–area relation using the bow force in ship–ice interaction. He defined this as a contact area given by the bow geometry, ice edge and ship motion, in accordance with the work of Riska. In fact, this definition is fully consistent with the global interaction area as defined in the present chapter. Figure 3.2 (b) illustrates the areas for a ridge interacting with a fixed structure. One can think of the global area as being fixed to the ice, but within the structure face, and the local area as being fixed to the structure.

A key factor in the design of offshore structures or individual structural components is the consideration of *exposure*. This is important both in terms of demand (load) and capacity (resistance or strength). In terms of structural load in a particular ice environment, which is our main concern, the exposure of a structure accounts for the number, severity and duration of ice interaction events. This can be measured in a number of different ways. For instance the exposure of an icebreaker might be measured by the number of rams into thick multiyear ice in a given time period (say, a year). More generally the exposure of an offshore structure in an ice environment could be considered as a function of the number of events such as interactions with large ice features per unit time. The size of the features could also be important; interaction with a floe that is 1000 m in diameter would result in a greater exposure than a floe with a diameter of only 100 m, since the duration of the impact with a 1000 m ice floe would be much longer than that of a 100 m ice floe, resulting in a greater likelihood of exceeding a particular load or pressure threshold. The longer the duration, the greater the likelihood of encountering higher pressures. The structure might be considered for two deployments at two different locations in the Beaufort Sea. At location 1, one might expect 5 impacts per year with multiyear ice floes, whereas at location 2, one might expect 20 impacts per year with multiyear ice floes. Here the number of impacts per year with multiyear ice could be considered a measure of the exposure of the structure.

Exposure of the structure can be dealt with in the context of probabilistic design. An example of the use of exposure and probabilistic methodology is given in Jordaan et al. (2014); see also Jordaan (2005). The Poisson distribution is useful in the analysis of series of collisions of ice with offshore structures or ships. Its use was first suggested by Kheisin (1973) in the context of ship maneuvering and by Jordaan et al. (1987, 1993b) for studying extreme ice loads. In the 1987 paper, an outline model for loads in arctic shipping was suggested. Kujala (1996) studied ice crushing in icebreaking as a Poisson process. The Poisson distribution is also used in ISO 19906:2019. Consider now the situation where there are n interactions in a time interval (for example, a year) with ice features in an area where a structure is located or where a vessel is operating. The Poisson distribution is given by

$$p_N(n) = \frac{\exp(-v) \, v^n}{n!}, \tag{3.2}$$

where v is the expected number of interactions. This process can be amended by factoring the arrival rate

$$v' = v \, [1 - F_X(z)], \tag{3.3}$$

where F_X is the cumulative distribution function for loads or pressures and Z is the extreme value (Jordaan, 2005). Then the extremal distribution is obtained from the value of the Poisson distribution using v' and with $n = 0$:

$$F_Z(z) = \exp\{-v' \, [1 - F_X(z)]\}. \tag{3.4}$$

This will be taken further in Section 3.3, using data from ship trials.

Table 3.1 Summary of ship-based data sets

Data Set	Years	Instrumentation	Data Available	Ice Type
CCGS *Louis S. St-Laurent* Voyage	1980	Pressure sensors and strain gauges	Local pressures	FY, SY, MY
Canmar Kigoriak	1981	Strain gauges on two instrumented areas	Local pressures	SY, MY
CCGS *Louis S. St-Laurent* Voyage	1994	Strain-gauged ship hull (bow, shoulder and bottom panels)	Local pressure and distribution	FY, MY
USCGC *Polar Sea* Voyages	1982 to 1986	Strain-gauged ship hull	Local pressures, global loads	FY, MY
Swedish I/B *Oden* Arctic Ocean Expedition	1991	Strain-gauged ship hull; accelerometers	Local pressures, vertical global loads	FY, MY
CCGS *Terry Fox* Bergy Bit Impact Study	2001	Strain-gauged ship hull; IMD Impact Panel; MOTAN system	Local pressures, global loads	Iceberg

3.2   Data from Ship Trials

The data sets analysed are summarized in Table 3.1. The analysis methods to be outlined were developed during the Review and Verification of Proposals for the Arctic Shipping Pollution Prevention Regulations (Carter et al., 1992), during which study the results of the *Canmar Kigoriak* trials were made available. This led to the analysis of the 1980 *Louis St-Laurent* voyage and the "Ice Data Analysis" (IDA) project (Section 3.5), in which several additional data sets were analysed, as summarized in Table 3.1.

3.3   The *Kigoriak* and the Alpha Method

The *Kigoriak* was the first commercial icebreaking vessel developed to support offshore oil exploration in the Beaufort Sea, with a displacement of 7,000 tonnes. The vessel was instrumented with 45 shear strain gauges and 7 strain gauges, and the loads on two instrumented areas A1 (1.25 m^2) and A2 (6.00 m^2) were measured. The data from the *Kigoriak* were collected during the August and October 1981 deployments of the vessel. The August tests were conducted primarily in thick first year and second year ice while the October tests were conducted in multi-year ice. Figure 3.3 shows the

Figure 3.3 The *Kigoriak* ramming thick ice. Source: Canadian Academy of Engineering. (Photo: Canmar).

Kigoriak ramming thick ice. During these tests 397 rams were recorded. The velocity of interactions varied from about 1 to 7 m s^{-1}. These were hard rams. More detail can be found in Dome Petroleum (1982). It should be noted that the Kigoriak design included a spoon bow, shaped for efficient breaking of ice sheets in flexure by applying a vertical force to the ice. If such a bow shape is used in thick ice, flexural failure does not occur, and very high loads are measured. The interaction does not involve pure crushing, rather crushing with sliding of the bow over the ice, which should be distinguished from crushing where the structure motion is predominately normal to the ice surface. An interesting analysis was presented in Dome Petroleum (1982), in which it is stated that normal forces develop between the vessel and the ice with a relatively slow normal interaction rate, resulting in relatively high pressures. This was deduced from the classical result of Michel (1978) in his "universal curve" for ice crushing and indentation (Figure 4.22 in his book). This showed that, at slow rates, corresponding to the "transition" range, the pressures were significantly higher than those at slower ("ductile") and those at higher ("brittle") ranges. This result is also reflected in later medium scale field indentation tests (Masterson et al., 1999). The Kigoriak results should be viewed as being conservative for design purposes. Bow shape is further addressed in Section 3.5.

The data sets comprised values of maximum force per ram. Rams can be considered as discrete events, as against interactions found in continuous icebreaking through an ice field (in the latter case, see also Kujala, 1991). The measure of exposure used in the present analysis is the ship ram; increasing the number of rams will increase the chance of a higher value of pressure. The two data sets were combined into a single set of rams, and the panel pressures ranked in descending order for each area. The probability of exceedance was calculated based on the ranked data, using the formula

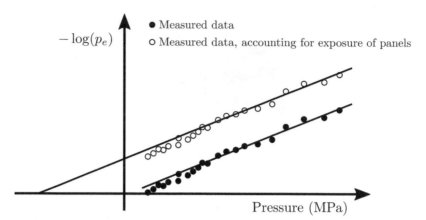

Figure 3.4 Exponential tail of ranked data plotted as an exceedance relationship.

$i/(n+1)$, where i is the rank and n is the number of rams producing pressure greater than zero on the instrumented panels.[1] There were 120 such rams for A2 and 181 for A1. For any given area, it was found that the extreme values of pressure tended to a linear relationship with $(\ln p_e)$, where p_e is the probability of exceedance. Figure 3.4 shows an example; further examples can be found in the parent publication (Jordaan et al., 1993a). As a result, the following relationship was used in the modelling, in which the key parameter "α" (in MPa) is defined:

$$p_e = \exp\left[-\frac{(x - x_0)}{\alpha}\right]. \tag{3.5}$$

In many cases (e.g. the *Polar Sea*, to be discussed in the sequel) there were sets of panels of identical area, and we have taken the maximum of all the pressures on a given panel area in a given ram. If there are j rams and m panels, the plotting position should be adjusted to $i/(m \cdot j + 1)$ to account for the correct exposure for a particular panel area. Figure 3.4 shows the correction for exposure of a set of panels of the same area. Studies were conducted to determine the best relationship; see Jordaan et al. (1993a) and Brown et al. (1996) for details. The deduced relationship for α (in MPa) is as follows:

$$\alpha = 1.25a^{-0.7}, \tag{3.6}$$

where a = area under consideration. This equation is plotted on Figure 3.9, and has been adopted in ISO 19906:2019. The results of the analysis were positive with regard to the review and verification of the Proposals for the Revision of the Arctic Shipping Pollution Prevention Regulations (see Carter et al., 1992) and confirmed the provisions for local pressure and design. The *Kigoriak* results turned out to be the most extreme of all the data sets analysed.

---

[1] This is the Weibull plotting position. For small samples, other plotting positions should be considered. See Fuglem et al. (2013).

3.4 Initial Modelling of *hpz*s: CCGS *Louis S. St-Laurent* 1980

The Canadian icebreaker CCGS *Louis S. St-Laurent* was used in two important trials: the first in 1980 and the second in 1994. The first of these became one of the
early demonstrations of the localization of ice pressure in compression. The second
trial will be reported in Section 3.5 as part of the IDA project. Blount et al. (1981)
and Glen and Blount (1984) reported on the 1980 Fall Arctic Probe into MY ice. The
principal objective of the trials was to measure the pressure generated in the impact
zone between the ship and ice. For this purpose, an array of 25 pressure sensors and
12 strain gauges were installed in the bow thruster compartment. The pressure sensors
were quite small (about 8 mm in diameter) so that the pressures measured represent
local pressures "at a point" quite well. During the ramming tests, five pressure sensors were inactive, reducing the contact height to 1.38 m and the "active window"
to 1.62 m². As an illustration of a typical result, Figure 3.5 shows a series of measurements taken during a ramming trial. The maximum pressure during the controlled
trials was 53 MPa; the maximum pressure in all tests was 70 MPa. The results show
that very high pressures can and do occur in ramming of ice features by icebreakers. It

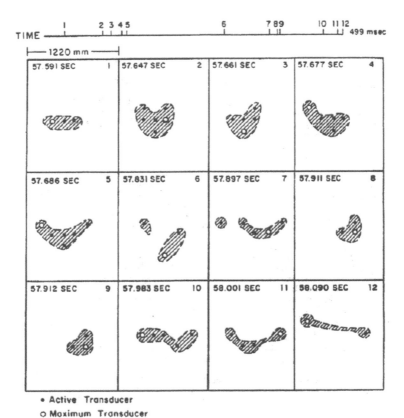

Figure 3.5 Occurrence of *hpz*s during a ram (Glen and Blount, 1984). With permission of
ASME.

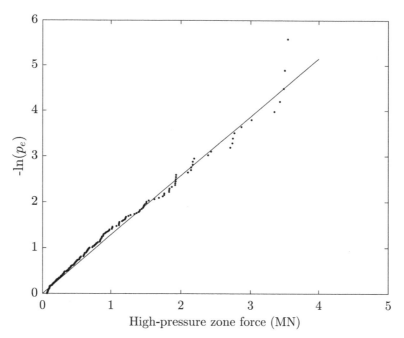

Figure 3.6 Results of Zou's analysis of *hpz* force magnitude.

should be noted that the spatial resolution is fixed by the arrangement of sensors. Glen and Blount (1984) describe typical normal velocities upon impact in the range up to 1.5 m s^{-1}.

Zou (1996) studied the "zonal forces" generated in the high-pressure zones using probabilistic methods. The pressure distribution over the instrumented panel was determined for various points in time and high-pressure zones were identified. For any instant in time, the *hpz*s can be identified and the associated force approximated as

$$Y = \sum_{i=1}^{n} y_i a_i, \tag{3.7}$$

where Y is the zonal force, y_i is the pressure on the ith pressure sensor with associated area a_i and n is the number of active pressure sensors. There is uncertainty related to the estimation of areas in the analysis and therefore two approaches were made regarding the areas represented by the sensors (termed assumptions A and B), which gave somewhat different *hpz* intensities, but which provided reasonable fits to exponential distributions. Peak pressures were selected at various points within the time duration of each ram, using on average 9 time slices per ram. The results were then ranked. Figure 3.6 shows the results for Assumption A (used also in Blount et al., 1981). There were approximately 1 *hpz* m^{-2} with an average force of 0.78 MN. The *hpz*s were approximately 0.1 m^2 in size; it would be expected that larger forces correspond to larger *hpz* areas.

This work suggests that the high-pressure zones might themselves be treated as being exponentially distributed, with a mean of 0.78 MN. This presents an opportunity to illustrate one factor contributing to the scale effect, probabilistic averaging

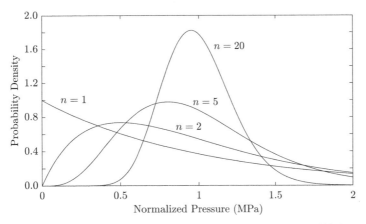

Figure 3.7 Averaging of *hpz*s over increasing area; n = number of high-pressure zones, for $\beta = 1$.

(Section 4.3). We consider first an area containing a single *hpz* with average pressure on the area X following the exponential distribution, and with mean pressure equal to β. The distribution of pressure is then given by

$$f_X(x) = \frac{1}{\beta} \exp\left(-\frac{x}{\beta}\right). \tag{3.8}$$

The parameter β also represents the standard deviation. If we consider a larger area, n times the original one, each containing a *hpz*, then we may derive the probability distribution of the average pressure, denoted as Y, where

$$Y = \frac{1}{n} \sum_{i=1}^{n} X_i. \tag{3.9}$$

On the assumption that the Xs are independent and identically distributed, Y is gamma-distributed:

$$f_Y(x) = \frac{x^{n-1}}{(n-1)!} \exp\left(-\frac{nx}{\beta}\right), \tag{3.10}$$

with mean β and a standard deviation of $\beta / \sqrt{n}$. Figure 3.7 illustrates the result for $\beta = 1$. See also Jordaan et al. (1993b). It can be seen that the distribution becomes more concentrated near the mean, and with large n, tends to the normal (Gaussian) form. This has indeed been observed for global forces (Kärnä and Qu, 2006). This figure also illustrates a scale effect with decreasing dispersion for large n, an effect that is largely statistical. Estimates of extreme values will decrease with increasing n.

A further analysis was carried out, in which a design area was loaded by a set of zonal forces, using the compound Poisson process. This was carried out using Monte Carlo simulation, with the design area being loaded by a set of *hpz*s, the number n in the set being determined by the Poisson process as a function of area a:

$$p_N(n) = \frac{\exp(-\rho a)(\rho a)^n}{n!}, \tag{3.11}$$

where ρ is the areal density (number per unit area) of hpzs, and a is the area. The zonal forces were modelled by equation (3.8) but for force rather than pressure, as independent and identically distributed quantities.

The approach just outlined is useful in design work; statistics can be developed for any of the IDA sets of rams and interactions, giving input parameters on hpz magnitude and areal density. These can be used in turn to formulate design pressures, including background pressures.

3.5   The IDA Project: More Results for the Alpha Curve

In the "Ice Data Analysis" (or "IDA") project (IDA 2007; Jordaan et al., 2007a), several additional sets of field measurements were analysed using the methods described in Section 3.3. The voyages were those by the CCGS *Louis S. St-Laurent* (1994 trial), the USCGC *Polar Sea* (8 trials), the Swedish Icebreaker *Oden* and the *Terry Fox* Bergy Bit experiment (Table 3.2). The data set is quite rich. Details of each vessel and the voyages are outlined in the following; further detail can be found in IDA (2007).

CCGS *Louis S. St-Laurent* (1994 trial). The vessel was built by Canadian Vickers in 1969. During her first two seasons, she escorted the USS Manhattan on both her historic Northwest Passage voyages. She is 120 m in length, with a gross tonnage of 11,345. An area 7.2 m long by 3 m high, 0.7 m below the waterline and approximately midway between the stem and shoulder was instrumented. During the summer of 1994, the CCGS *Louis St-Laurent*, in tandem with the USCGC *Polar Sea*, transited the North Pole (Ritch et al., 1999; Frederking, 1999). The two vessels left Alaska (Chukchi Sea), crossed the North Pole and arrived at Svalbard (Fram Strait). During the voyage, a total of 1730 bow impact events were recorded, with an average speed of 1.9 m s^{-1}. High ice concentrations resulted in the need to operate in a continuous icebreaking or ramming mode.

USCGC *Polar Sea* (8 trials). The USCGC *Polar Sea* was built by the Lockheed Shipbuilding and Construction Company in Seattle, Washington, in 1978. She is 122 m in length, with a displacement of 11,000 t. A 9.1 m^2 area of the bow was instrumented

Table 3.2 Ice pressure measurements on USCGC Polar Sea

Location	Measurement Dates	Ice Type	Number of Impacts
Alaskan Beaufort Sea	28 Sept–16 Oct 1982	MY	167
South Bering Sea	24–26 March 1983	FY	173
North Bering Sea	27–28 March 1983	FY	243
South Chukchi Sea	29 Mar–2 Apr, 28 Apr–2 May 1983	Mixed FY and MY	299
North Chukchi Sea	3–27 April 1983	Mixed FY and MY	513
McMurdo Sound, Antarctica	9–13 January 1984	FY	309
Beaufort and Chukchi Seas	18 Nov–1 Dec 1984	Mixed FY and MY	337
Bering Sea Ice Edge	18–27 April 1986	FY	653

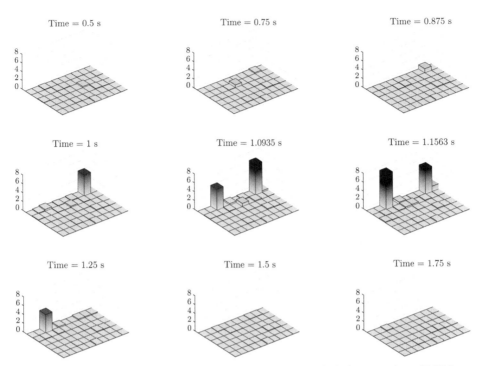

Figure 3.8 Pressure measurements in MPa on the *Polar Sea* panels during ramming of MY ice, showing the appearance and disappearance of *hpz*s. Beaufort Sea, 14 October 1982.

with strain gauges. The instrumented panel was divided into 60 sub-panels, each having an area of 0.15 m^2. The relatively small size of the sub-panels makes the *Polar Sea* data suitable for high-pressure zone investigation, given an average size of an *hpz* of about 0.10 m^2. The data collected has been used for local pressure extremal analysis and for the analysis of high-pressure zones. The USCGC *Polar Sea* made eight different voyages during which data were collected between 1982 and 1986, amounting to over 2,600 ice loading events. Reference is made to Daley et al. (1984) and St. John et al. (1984) for information on the test programmes. Table 3.2 summarizes the voyages. Speeds were generally up to 6 m s^{-1} for the 1982 and 1983 FY and MY events. Figure 3.8 shows typical results of measured pressures on panels of the Polar Sea during ramming. This shows in a clear fashion the appearance and disappearance of high-pressure zones.

Oden. The Swedish icebreaker *Oden* was built in 1988 and has a displacement of 13,000 t. The IB Oden was delivered by the Götaverken Arendal shipyards to the Swedish Maritime Administration (SMA) in 1989. The *Oden* was part of a three vessel expedition to the central Arctic Basin, the International Arctic Ocean Expedition 1991, to conduct a broad range of scientific measurements in the Arctic Ocean. The expedition left Tromso, Norway, and followed a track to the west of Spitzbergen, Franz Josef Land, the Lomonosov Ridge, and the Makarov Basin, ultimately reaching the North Pole. Strain gauges were used to estimate ice pressures. The instrumented panel was divided into 32 sub-panels, each with an area of 0.65 m^2. Data from these events were used for local ice pressure analysis. Ship speeds covered a large range, up to over

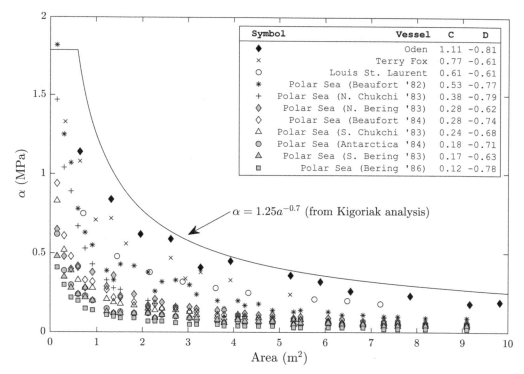

Figure 3.9 Curves of α versus area for the IDA set of vessels. The curve deduced from the *Kigoriak* trials is also shown. With permission of ASME.

8 m s^{-1} See also St. John and Minnick (1993). It should be noted that the *Oden* was designed with a very flat "spoon bow" (as was the case for the *Kigoriak*).

***Terry Fox* Bergy Bit Experiment.** The CCGS *Terry Fox* was built in 1983 in Victoria, British Columbia, by Burrard Yarrows Corporation. Gulf Canada had the Terry Fox and her sister ship the Arctic Kalvik designed and built for multi-purpose heavy tug and supply vessel operations in the Beaufort Sea. In 1993, the ship was purchased for the Canadian Coast Guard from Gulf Canada and stationed at Dartmouth, Nova Scotia. The instrumented area covered 5.4 m^2, with a minimum panel size of 0.08 m^2, in a strain-gauged area on the ship bow. The typical areas suitable for design purposes derived from the three differently shaped areas were 0.24 m^2, 0.12 m^2 and 0.08 m^2. The experiment consisted of ramming of icebergs, with 178 impacts, 60% of which caused loads on instrumented area. There were repeat loadings on icebergs, with 18 different iceberg masses tested, relatively small icebergs, up to 8500 tonnes. Speeds were in the range of 2 to 7 m s^{-1}. The programme was reported by Ritch et al. (2008), who also gave an analysis similar in method to that described in IDA and in the present work.

Analysis of IDA Voyages. The methodology follows that outlined in Section 3.3, based on equation (3.5), developed for different areas, with associated extremal analysis (see Figure 3.4). The results of the analysis are presented in Figure 3.9. The alpha-curves follow a remarkably close relationship with a power-law decrease of the form (area)$^{-0.7}$. Individual values of coefficients in the power-law relationship Ca^{-D}

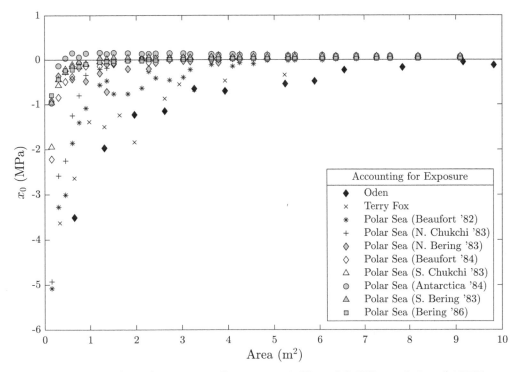

Figure 3.10 Values of x_0 corresponding to curves in Figure 3.9. With permission of ASME.

are given in Figure 3.9, with values of x_0 in Figure 3.10. Results were reported in IDA (2007), and also noted in Jordaan et al. (2007a); the team responsible for conducting the analyses is noted in these reports. The results were later summarized in Taylor et al. (2010). We have generally assumed that $x_0 = 0$ in equations (3.5), (3.17) and (3.20). This is generally a good assumption but should be checked. Equation (3.20) shows that if it is negative, it reduces the design pressures, but if positive, it adds to design requirement. Often small negative values have been found.

Equation (3.6) has been written more generally as

$$\alpha = B_p a^{-0.7}, \tag{3.12}$$

and appears in ISO 19906:2019.

Application in Design. The results described can be used for design purposes. We can consider the potential occurrence of m events (say, rams) in the design period of time (say, a year), and we wish to obtain the distribution of the maximum Z:

$$Z = \max (X_1, X_2, \ldots, X_m). \tag{3.13}$$

Then

$$F_Z (z) = F_X^m (z), \tag{3.14}$$

for independent and identically distributed (*iid*) quantities. We now assume that the occurrence of rams can be modelled as a Poisson process as discussed in equation (3.2), and that the expected number of events is μ. Then, using equation (3.4),

$$F_Z(z) = \exp\{-\mu[1 - F_X(z)]\}, \qquad (3.15)$$

and, using equation (3.5), we have

$$F_Z(z) = \exp\left\{-\mu \exp\left[\frac{z - x_0}{\alpha}\right]\right\}. \qquad (3.16)$$

We may consider "hits" and "misses" on a particular panel using $\mu = r\nu$, where $\nu =$ number of interactions with $r =$ proportion on "hits". Then we can write

$$F_Z(z) = \exp\left\{-\exp\left[-\frac{z - x_0 - x_1}{\alpha}\right]\right\}, \qquad (3.17)$$

where

$$x_1 = \alpha \ln \mu = \alpha(\ln \nu + \ln r). \qquad (3.18)$$

Many projects are mission-oriented. We might be considering the impacts of ice features with an FPSO or with a fixed offshore installation, at a certain location. The duration of exposure in the interaction under consideration, t, can be calculated from the interaction mechanics, and compared to t_{ref}, the duration of the impact phase for the reference data set, e.g. for the Polar Sea or other vessel used in the analysis. Given η, the frequency of impacts per year, one can arrive at the value μ for the design under consideration:

$$\mu = \eta \frac{t}{t_{ref}}. \qquad (3.19)$$

We may then consider the extreme value z_e for use in design, with an exceedance probability $[1 - F_Z(z_e)]$. For example, for the 1% exceedance value, $F_Z(z_e) = 0.99$. Then, from equation (3.17):

$$z_e = x_0 + \alpha\{-\ln[-\ln F_Z(z_e)] + \ln \nu + \ln r\}. \qquad (3.20)$$

In many cases $x_0 \simeq 0$, and we may write, to a close approximation, that

$$z_e \simeq \alpha(\ln n + \ln \mu), \qquad (3.21)$$

where n is the return period under consideration. This is a useful result for practical applications and has been applied in several studies.

Velocity Effects. In considering the effect of velocity on ice pressures and forces, it is often taken as axiomatic that higher speeds of collision lead to higher loads and pressures. This is not necessarily the case: individual campaigns show that pressure and force are largely independent of the velocity of impact. Some results are shown in Figures 3.11 based on Dome Petroleum (1982), St. John et al. (1984) and Daley et al. (1986). There is some indication of an increase in force or pressure up to 2 m s^{-1}, but no trend thereafter. This result had been noted by Tunik (1991). Although there is in effect little or no dependence of pressure and force on velocity in any particular trial, it must be emphasized that different field trials with different conditions result in different severities, as shown in Figure 3.9.

Discussion and Bow Form. In using the data, the conditions envisaged for the operating area of the installation or vessel under consideration should be matched with

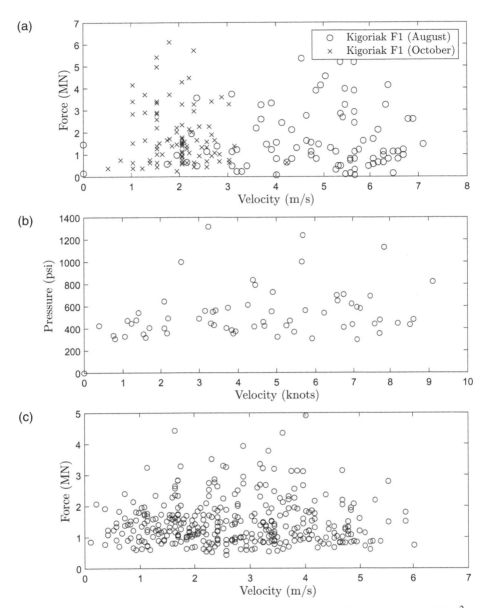

Figure 3.11 Effect of velocity on pressures and forces: (a) *Kigoriak* (1981) forces on 1.25 m^2 area; (b) single panel pressures in multiyear ice, *Polar Sea* 1983; and (c) single panel forces on the *Polar Sea*, North Chukchi Sea 1983.

the conditions under which the measurements were made. There are a large variety of ice conditions and ramming speeds. The geometry of the bow will play a role in this; in particular we mention the spoon bows of the *Kigoriak* and the *Oden*. These are designed for efficient breaking of ice sheets in flexure. If such a bow shape is used in thick ice, flexural failure does not occur so that very high loads result. The *Kigoriak* ramming was in thick ice and the *Oden* also encountered some very thick ice, up to

6 m (Frederking, 2005). The results from these two trials are the most severe of those presented in Figure 3.9. It is therefore judged that these curves are conservative for many applications. Yet the *Kigoriak* data have served well during the years before the full range of data sets became available. It is judged that the values from vessels with spoon bows used in thick ice exceed those from crushing where the structure motion is predominately normal to the ice surface. For another discussion of the ice pressures, see Frederking (2005) and Frederking and Kubat (2007).

Applications to design are described in Section 5.3. A related method is based on the upcrossing rate, developed by Li et al. (2010), whereby the rates of exceeding specified levels of pressure or force are determined. This is a very useful method and avoids the consideration of particular events, such as ship rams. This is a good approach when considering vessels transiting an icefield, or when the ice field moves past a moored vessel. The event-based method was discussed in Manuel et al. (2016) in a study in which this was contrasted with the upcrossing rate method in Ralph et al. (2016).

4 Fixed Structures and Medium-Scale Indentation Tests

4.1 Introduction

Generally the concern in this chapter is with wide structures in ice, although tubular structures have been used in areas such as Cook Inlet. In the case of fixed structures, the ice speed is generally less than 1 m s^{-1}, although a collision of a bergy bit with a moored floater might be affected by the current sea state, with waves accelerating the ice piece. In this case, speeds of several metres per second are possible so that data from ship trials are more relevant as a basis for design. Grounded fixed structures will possibly experience ice-induced vibrations, while responding elastically to load. The possible formation of *hpz*s is relevant; very slowly moving ice will impart a relatively uniform distribution, but faster rates result in *hpz* formation and possible dynamic actions.

Much of the content of this chapter follows from imaginative experiments over the last several decades. These include groundbreaking measurements made in the Canadian frontier regions: the Hans Island experiments, the Molikpaq experience and the medium scale indentation series of tests. Lessons have been learned, especially from the Molikpaq, with regard to ice-induced vibration.

4.2 Helsinki Tests, JOIA and STRICE

Riska (1991) conducted experiments of structures interacting with ice in the laboratory (up to 15 cm s^{-1}) on specially prepared ice samples and in the field on level ice. The latter involved a moving vessel, but were similar in concept to the laboratory tests. The contact was observed visually and contact pressures measured with pressure-sensitive film (PVDF film). The main observation was that the contact is concentrated on a narrow line-like region where the pressure is high locally, up to 50 MPa. The results were verified in full scale, with tests on the icebreaker Sampo, using a Lexan window and pressure-sensitive film, in level ice of thickness 40 to 60 cm. A similar set-up to that used in the laboratory was used to measure contact ice pressure. The window in the contact zone at the waterline was constructed so as to observe ice contact. Observations showed again the contact as being in a narrow line, in fact a sequence of *hpz*s with pressure up to 50 MPa on small areas of size 2.25 cm^2 within the *hpz*s being measured. Calculation of the correlation between the individual pressure sensor signals shows correlation between sensors at different waterline levels is small (< 0.25) but also that the correlation is small also along the line.

The Japanese Ocean Industry Association (JOIA) conducted the Medium Scale Field Indentation Test (MSFIT) programmes during the winters of 1998, 1999 and 2000; these test results are commonly known as the JOIA data (not to be confused with the later Medium Scale Indentation Tests discussed below). The scale of the indenter used in these tests was relatively small (1.5 m in lateral dimension). For background, reference is made to Takeuchi et al. (1997, 2000) and Akagawa et al. (2000). Frederking (2004) summarized some test results while Taylor and Jordaan (2011a) summarized the programme and used the results to study probabilistic averaging (Section 4.3). The tests were carried out over a number of winters in sea ice in the harbour of Notoro Lagoon in Hokkaido, Japan. The indentor apparatus was mounted in a testing frame that allowed the unit to be deployed from the side of the fishing dock. The instrumented indentor consisted of a 1.5 m beam segmented into fifteen panels. Each segmented panel was 10 cm wide by 40 cm high and was fitted with a 10 metric ton load cell to measure local panel loads. In addition to the segmented panels, tactile sensors were fitted to the indenter for some experiments. The average ice thickness during these tests was of the order of 30 cm. For some tests naturally grown ice was used, while in other cases the natural ice was removed and a refrozen ice sheet was grown in the test region. Loading was servo-controlled at a constant displacement rate in a range from 0.03 cm s^{-1} to 3 cm s^{-1} using a hydraulic ram controlled by a power pack unit.

The programme provided important new information and discoveries. First, the extreme non-uniformity of the load distribution that the ice applies to the indentor was quantified, reinforcing the results from the Helsinki tests. Akagawa et al. (2000) discussed sample cases of tactile pressure sensor and segmented indenter data regarding peak loads and the failure modes observed. The results support the concepts of "line-like" loading and independent failure zones (high-pressure zones) along the "lines". At the higher rates, the stressed area could be as low as 10% of the nominal area, and appeared as a line-like set of *hpz*s, along the central region of the interaction area. The failure mode of the ice changed from a more ductile mode with distributed pressure at slower speeds to brittle crushing with the line-like distribution at higher speeds. Frederking (2004) discussed a test conducted under the following conditions: Indentor width = 1500 mm, Indentor height = 500 mm, Ice thickness = 168 mm, Indenter speed = 3 mm s^{-1}. Figure 4.1 shows the result at the 65 s mark with two successive tactile film records: one just before failure, "peak", followed by one just after failure, in the "trough". The loads at the two points in time, estimated from the tactile film, were 85 kN and 53 kN, respectively. The data are particularly suited to the evaluation of probabilistic averaging methodology since these tests provide both local and global pressure data for the test structure. This was treated in Ian Jordaan and Associates (2007) and by Taylor and Jordaan (2011a), in which autoregressive methods were used, including a bilinear analysis. The line-like nature of loads exerted by ice sheets failing in crushing has been recognized in several other studies; see for example Adams et al. (2019).

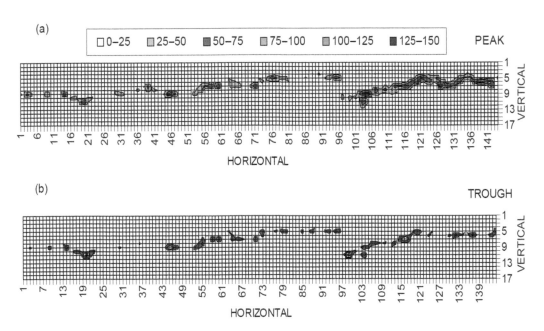

Figure 4.1 Peak and trough contour plots of local pressure in JOIA data (Frederking, 2004). With permission from IAHR.

Full scale measurements of ice forces were carried out at the Norstrømsgrund light-house as part of the European Union funded projects LOLEIF and STRICE from 1999 to 2003. These are described in Kärnä and Qu (2006). The lighthouse is located in the northern part of the Gulf of Bothnia where only first year ice is encountered and where water salinity is about 1 part per thousand (the ocean's least saline seawater). It is founded at a water depth of 14 m and has a waterline diameter of approximately 7.2 m. The lighthouse was outfitted with 9 large surface (1.65 m × 1.25 m) ice force measuring panels, with the arrangement as in Figure 4.2 (from Taylor, 2010). The force panels used in the LOLEIF and STRICE projects were 1.2 m wide and the ice thick-ness varied from 0.2 m to about 1.0 m. The panels were mounted half way around the lighthouse (covering 162° of the perimeter, where each panel covers 18° of the perimeter) directed to ENE as shown in Figure 4.2. The ice thickness was measured by a sonar device which was mounted at the bottom of the lighthouse foundation at 7 m water depth several meters in front of the vertical shaft. As a redundancy for the sonar ice thickness measuring system, a laser distance metre was installed in 1999 hanging about 2.0 m above the ice in front of the lighthouse.

Kärnä and Qu (2006) summarized the results of 215 events of stationary ice crush-ing events occurring at the lighthouse from 2000 to 2003. The mean ice forces and associated parameters were obtained by combining data for brittle and low velocity crushing events, respectively. These are given in Tables 1.1 to 1.6 and 2.1 to 2.6 of the report. The mean ice pressure and mean standard deviation for each event were reported in a usable form. These are very useful results that can be used in design analysis. The correlation between adjacent panels is needed for global analysis. Guo

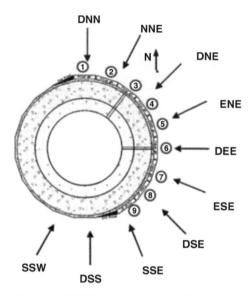

Figure 4.2 STRICE panels and load directions.

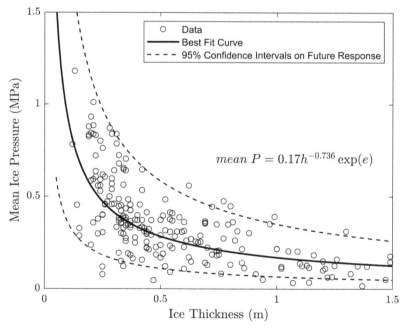

Figure 4.3 Pressure-thickness relationship (based on STRICE data).

(2012) found for crushing events that the correlation coefficient at the distance 1.2 m (one load panel width on the lighthouse) is about 0.25, whereas the coefficient at the distance 2.4 m (twice the load panel width on the lighthouse) is less than 0.1. The data have been used for the determination of probabilistic global and local loads on structures in the Kashagan Field, in the northeast Caspian Sea (Jordaan et al., 2011b). This is also a low salinity environment, and the paper describes the modelling of interaction events and the use of STRICE data. The STRICE results showed a strong size effect as shown in Figure 4.3. These data were used in the analyses just described to obtain design pressures.

4.3   Probabilistic Averaging

We have discussed in Section 3.4 the idea of averaging independent *hpz* forces to obtain loads on larger areas. Figure 3.7 shows the result: the standard deviation of the pressure reduces by the factor $(1/\sqrt{n})$, where n is the number of *hpz*s. This was based on the assumption of independent and identically distributed *hpz* forces. Kry (1978, 1980) quite early in the development of ice mechanics suggested that the total load on a structure might be treated as the sum of failure of independent zones, and that probabilistic analysis shows that peak loads over a larger area are less than the sum of maxima on individual sub-areas. His focus was on wide structures. The work of Ashby et al. (1986) on nonsimultaneous failure is also pertinent as is the work of Kujala (1996), who modelled ice crushing as a random Poisson process.

An application where local correlations between adjacent areas were used is given in Jordaan et al. (2006, 2007b), using data from the Molikpaq structure. Specifics of instrumentation are given in Section 4.4, in which a detailed evaluation of the measurement systems is made; for the present purposes, reference is made to Figure 4.6, which shows the set of "Medof" panels. These were designed to measure loads locally; the total load on the structure might be based on the average pressures on individual Medof panels. The problem being addressed initially was that the global loads estimated from the Medof panels were significantly higher than those estimated from other readings and inferences. This is addressed extensively in Section 4.4; for the present we discuss only the question of local correlations: some Medof panels were adjacent, others were distant from each other. Analysis of data showed that loads on adjacent Medof panels had correlation coefficients of between 0.6 and 0.8, whereas distant ones were uncorrelated. The analysis is aimed at showing the effect of the correlation between adjacent measurements on global load estimation. Figure 4.4 shows the main result. The averaged pressures show a much wider distribution than the distribution in which the correlation has been taken into account. The assumption of independence between zones results in overestimates of global loads. We shall see in Section 4.4 that the averaging was not sufficient to explain the differences between the estimates based on different measurement systems.

If data on global loads on a structure result from direct measurement thereof, there is of course no need for any correction, and this includes the case where pressure panels cover the entire area under consideration. In the case where the panels cover only part of the face, a correction should be applied for adjacent panels. There are several ways to account for the correlation. In the case under discussion, autoregressive analysis (Vanmarcke, 1983) was used, essentially a Markovian assumption in space. An interesting aspect is that averaging results in the normal distribution provided the

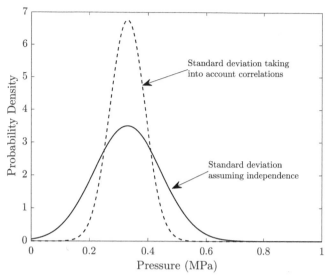

Figure 4.4 Effect of probabilistic averaging over panels on average pressures for Molikpaq structure.

conditions of the central limit theorem apply. A broad statement of the central limit theorem is that if one adds together or averages many random quantities, one obtains a random quantity that approaches the normal distribution. This holds for independent and identically distributed random quantities with finite variances. It has also been proved for less demanding conditions, for instance where there is some dependence between the random quantities, provided that the correlation is small except for that between one quantity and a limited number of others. This tendency to a normal distribution for global loads was found in measured data by Kärnä and Qu (2006).

4.4   Molikpaq

The Molikpaq is a mobile arctic caisson (MAC), commissioned by Gulf Canada Resources Ltd., and designed and built in the early 1980s as an exploration structure for use in the Beaufort Sea. It was extensively instrumented to measure its response to ice loading. The data from measurements at Amauligak I-65 (70°04′ N and 133°48′ W) in 1985–86 were used in a joint industry project (JIP) conducted by Gulf Canada Resources Ltd., and were reported in a series of confidential reports produced over the period 1987–91 (the 1987 JIP). ConocoPhillips Canada again made the data available in 2007 for a new study, the Ice Load Re-analysis (the 2007 JIP), which focused on multiyear ice loading on the Molikpaq at Amauligak I-65. The structure is illustrated in Figure 4.5, which shows an interaction with an ice floe, and the production of piles of crushed ice associated with *hpz* occurrence. Jordaan et al. (2018) provide background information on the original 1987 JIP conducted by Gulf Canada Resources Ltd., as well as the later re-evaluation (the 2007 JIP). A detailed summary is

Figure 4.5 The Molikpaq structure under ice loading. Source: Canadian Academy of Engineering. (Photo: G. Comfort).

given in Frederking et al. (2010, 2011); see also Jordaan et al. (2011) and Ian Jordaan and Associates (2010). Reports generated during the 2007 JIP can be found on the website www.engr.mun.ca/~2007JIPonMolikpaq/. The main objective was to investigate the reasons for the discrepancy observed at the time the 2007 JIP was initiated between loads from the Medof panels, as discussed below, and other measurements and estimations.

The paper by Jordaan et al. (2018) summarizes the details of the structure and associated instrumentation, as well as the JIP itself. The caisson has near vertical sides (7° from the vertical for the Amauligak I-65 location at the waterline). The caisson is a chamfered square in plan. At the waterline, the long sides are about 58 m long and the cropped corners 22 m long. When deployed, the minimum waterline width of the caisson was 90 m, and the maximum width across the diagonal was 104 m. The box girder deck rested upon the main structure, a steel annulus, with rubber bearings between the deck and the structure. For the 1985–86 deployment of the Molikpaq at Amauligak I-65, water depth was 31 m and the caisson was set down on a submarine berm, giving it a draft of 19.5 m. Ice loading events have been summarized in Hardy et al. (1996).

A brief outline of the 2007 JIP is now given. The core of the structure was filled with hydraulically pumped sand, uncompacted. The in situ state of core sand, being loose, was a key factor in initiating the re-evaluation, as summarized in Hewitt (2010). Since the sand was loose, significant mobilization of base shear under lateral load occurred before the sand core was activated. This led to lower ice load estimates by Hewitt than had previously been made in the literature so that considerable uncertainty

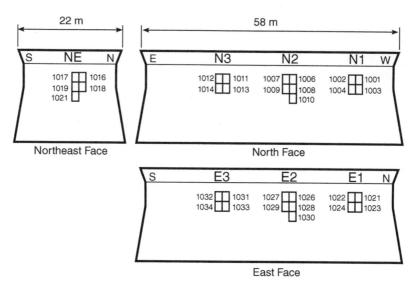

Figure 4.6 Arrangement of the Medof panels. With permission from Canadian Science Publishing.

was associated with the estimates. The main focus of the 2007 JIP was to investigate the accuracy of the measuring systems for ice loads. These included strain gauges, extensometers, and Medof panels. There had been for some time discussion over a discrepancy between the loads indicated by extensometers vs. Medof panel readings; see for example Sandwell (1991). The extensometers measured the axial deformation between the box girder deck and the steel structure, in effect providing estimates of the deformation of the structure itself. The Molikpaq acts essentially as an elastic "proving ring". The Medof panels (1.135 m wide and 2.715 m high) were composed of two parallel steel plates with Adiprene L100 urethane buttons sandwiched between the plates. These panels were designed to measure ice loads, and the arrangement is shown in Figure 4.6. The buttons covered a little less than half of the panel area. The remaining space between the steel plates was filled with liquid calcium chloride solution. The response of each Medof panel to loading was measured in terms of the height of the calcium chloride solution in a measuring tube. The polyurethane buttons were intended to act as "springs" to be calibrated and used to measure load by their deformation.

The geotechnical inferences, and the estimates using the extensometers, suggested global loads significantly lower than the estimates from the Medof panels, which had been used as the main source of load estimation. It was first thought that probabilistic averaging (Section 4.3) might explain the difference between the global loads measured by the extensometers, and the local ones measured by the Medof panels, but this averaging did not explain the large discrepancy. The original calibrations of the Medof panels were used to estimate ice loads. Calibration had been carried out using uniform pressures up to 1.86 MPa. The actual loading in the field of the Medof panels was extremely non-uniform spatially, particularly during ice crushing, and is very different from that used in the original calibration. There were loading incidents involving crushing and "pounding" of the structure by the ice (see also Section 4.8), including several events at the the previous location (Tarsuit P-45) the year before. It is now known that *hpz*s occur during ice crushing with local pressures on small areas of the order of 50 MPa or more. The resulting high pressures led to softening of the polyurethane buttons in the panels, including the Mullins effect. The panels were consequently reading too high. Calculations regarding a decelerating floe impacting the structure in open water conditions on 12 May 1986 were made (Fuglem et al., 2011). These calculations resulted in load estimates that supported those using the extensometers.

The study ended by supporting the conclusion that global ice loads were about half those obtained using the original Medof panel calibrations. Extensometer readings were seen as representing reasonable estimates of ice loads, with some uncertainty regarding the estimate of the load distortion ratio of the "proving ring". This method is used in estimating the loads on the 58 m faces of the Molikpaq during some of the key events (reported in Ian Jordaan and Associates (2010) and in Jordaan et al. (2018)) as part of the 2007 JIP. A best-estimate load distortion ratio of 2.60 MN mm^{-1} in the extensometers was used in deriving the pressures. The results are presented in Figure 4.7 (a). It should also be noted that the buttons in the Medof panels do not capture details of dynamic response found in some interactions. For example, Frederking (personal communication) suggests that a response up to 0.5 Hz is determined. The main conclusion of the work is that all past estimates of load or pressure based on the Medof panels should be reconsidered in the light of the study findings, and that grounds exist to consider a 50% reduction in the value of load.

A study of the local pressures measured on the Medof panels was undertaken by Jordaan et al. (2010). This work was completed before the re-evaluation of the ice loads (Jordaan et al., 2018) had been completed, so that the results in the 2010 paper do not reflect the conclusions noted above with regard to the Medof panels and associated softening. This correction will be discussed in the following. The Medof panels cover an area approximately equal to 10% of the area of a face during interaction with ice. The analysis focused on second year and multiyear ice interactions. A global interaction event is defined for present purposes as the load or average pressure on the structure as a whole during the time that the ice continuously presents a load, say an hour's duration. The local pressures will vary during this global event, sometimes diminishing to zero, but applying significant pressures lasting several minutes during the global duration of one hour. The load passes from one area along the loaded face to another. The determination of a local pressure event was based on the Medof pressure exceeding 0.125 MPa in adjusted measure, i.e. adjusted to reflect the discussion in the preceding paragraph (0.25 MPa unadjusted). This then defines a "local event". There can be more than one local event per global interaction event. The average duration for local events was found to be approximately 10 min with a standard deviation about 10 min, the results fitting an exponential distribution (see Jordaan et al., 2010, for details). The average duration of the global interaction events was about 52 min.

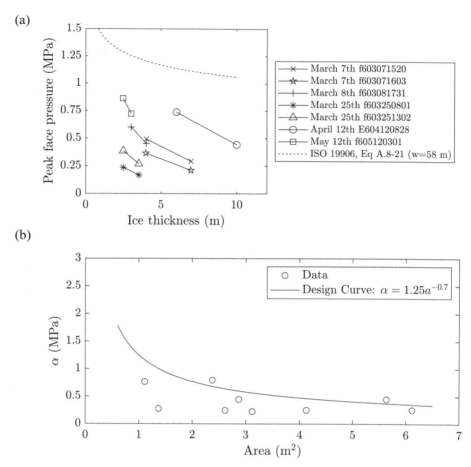

Figure 4.7 Results of analysis of pressures on the Molikpaq. (a) Best estimates of peak face pressures for some of the key events, based on extensometer readings. Pairs of points connected represent an upper and lower estimate of the ice thickness used to determine the peak face pressure. Numbers in legend refer to event number; w is the structure width in the ISO equation noted. (b) Alpha versus area for the Molikpaq; data have been adjusted to reflect panel softening. With permission from Canadian Science Publishing.

Sets of loaded areas for various load cases were formulated by taking into account panel combinations and ice thicknesses; details can be found in Jordaan et al. (2010). The values of α and x_0 (see equation (3.5) and Section 3.5) were determined for each area using the same method as before by taking the maximum pressure in each case. Figure 4.7 (b) shows the result, but with the values adjusted from the 2010 paper to reflect panel softening. It is seen that the use of the relationship $\alpha = 1.25a^{-0.7}$ gives somewhat conservative results. The average value of x_0 was found to be small (0.3 MPa), and therefore the assumption $x_0 = 0$ is reasonable. The analysis shows the important result that an exposure of a local area for 10 min of interaction with the Molikpaq structure is approximately equivalent to one ship ram of the Kigoriak (Section 3.3). The remaining figures in Jordaan et al. (2010) can be used despite the correction noted regarding the Medof panels; the value of α is still acceptable, but now somewhat conservative. Data used in the analysis shows that there is on average approximately one local interaction event per panel per global interaction event.

The loads estimated from the Molikpaq structure have been used in codes and standards. One issue to be resolved is the extrapolation from the events measured on the Molikpaq during a single season to the Extreme-level (EL) and Abnormal-level (AL) cases specified in ISO 19906:2019. The designer thinks in terms of events per year. Such an analysis of exposure will give different results for different geographical areas with different exposures. Based on the local ice pressure analysis just discussed, approximate estimates of increases in design load at the 1% annual exceedance level (EL load) are estimated approximately as 50% greater for 10 multiyear floe interactions per year compared to a single event, 85% for 50 interactions and 100% for 100 interactions. These estimates can be used together with Figure 4.7 to make comparisons to specifications in ISO 19906.

4.5 Hans Island

A highly imaginative test programme to obtain measurements of global pressures on large areas was performed in 1980, 1981 and 1983. The test programme was conceived by Metge and Croasdale. The experiments were carried out on a small rocky island between Ellesmere Island and Greenland, named Hans Island. It had been observed during overflights that thick multiyear ice floes were impacting the island at quite high speeds. Metge (1994) provides a very good summary of the three years of testing; Sanderson (1988) deals with the first two years. Both give good background descriptions. The edge of the island was covered by an "ice foot", a large grounded rubble pile. The edge is nearly vertical, providing an excellent contact surface for studying crushing failures. Occasionally, there were ice pieces between the impacting floe and the island and the impacts were deemed to be "cushioned"; in other cases a "clean" impact was found. The deceleration of the floes was based on measurement by several types of instrumentation including accelerometers, theodolites, electro-optical distance measurements and photogrammetry; thickness was based on auger holes and surveys of freeboard. Area was also surveyed.

We shall focus on one impact, floe 10 in the 1983 series. The floe was 5.2 m thick with a mass of 3.5×10^{10} kg and an initial speed of 0.87 m s^{-1} (Metge, 1994). Figure 4.8 shows the multiyear ice floe impacting the island, with a large fracture developing. My colleague, John Fitzpatrick, who was present at the event, has noted in a private communication "I remember the floe 10 experience very well. The interaction lasted maybe about 5 minutes. The floe had some melt pools on its surface which perhaps gave the illusion of 'weakness'. Not true at all. As the ice crushed there was tremendous noise, shuddering and shaking of the ice foot against which it crushed. When the floe eventually stopped, thousands of tonnes of fine, blue, powdered, crushed ice, floated and bubbled, almost 'frothed' up on the sea surface. Really quite amazing."

Figure 4.8 Floe 10 impacting Hans Island. Source: Canadian Academy of Engineering. (Photo: Michel Metge).

The force applied to the island was estimated based on measurements of the deceleration of the floe in question, its mass having been obtained from areal extent and thickness together with density. Based on measurements, John Fitzpatrick relates that he and fellow engineers estimated the force that evening on the island at 45,000 tonnes (about 450 MN). At the time, the estimate was so low (based on knowledge of the time) that they thought that a mistake had been made. Metge (1994) gives good details of the various impacts and associated sizes, speeds, forces, masses contact areas and related pressures. For our example case, the 1983 floe 10, the peak average pressures were 300 and 233 kPa (2 s and 10 s average respectively). The global pressures were, as noted, much lower than had been expected at the time. Sanderson (1988) relates that, at the time of the experiments, experience from artificial islands in the Beaufort Sea was limited to slowly moving first year ice, with pressures less that 1 MPa. There was concern that thicker, fast-moving ice might lead to much greater loads. The Hans Island did much to change the estimation of ice loads under such conditions. A further conclusion is that comparisons with the Molikpaq pressures have fallen into line more closely as a result of the re-evaluation discussed in Section 4.4.

4.6   Size Effect of Pressure and Ice Thickness

We have already seen in Figure 4.3 that there is a strong pressure-thickness relationship found in the STRICE data. The pressures for different ice thicknesses were studied in the context of the Molikpaq results in Taylor and Jordaan (2011b), noting that the width of the STRICE and Medof panels are similar. The results are shown in Figure 4.9. In this, the Medof panel results have been adjusted by a factor of 0.50 as proposed in Section 4.4. Events of duration of less than 10 min have not been included. Both first-year and multi-year ice were included from the Molikpaq data set to allow for a broad range of ice thickness values. The figure gives further evidence supporting

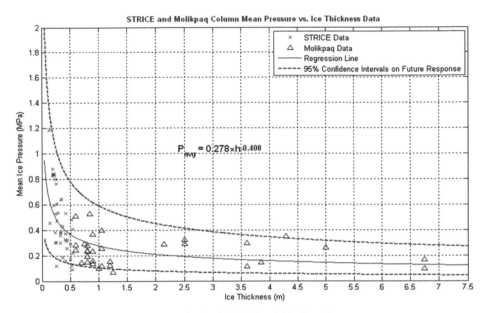

Figure 4.9 Mean pressure-area data for the STRICE and Molikaq data sets.

the decision to adjust the Medof panel results in the manner indicated in Section 4.4. In the absence of this adjustment, there would have been a misalignment of data, with a seeming "jump" to higher-than-average pressures, with no reasonable explanation. This conclusion is further supported by the results of the Hans Island experiments, which indicate pressures consistent with the adjustment of the Medof panel values.

It is important to note that the pressure–thickness relationships discussed hold for the faster interactions with crushing-type failures. For slow creep-type failure without fractures, pressures would not be expected to show this dependence on thickness or area. See the results of Li et al. (2004), which are further discussed in Section 8.7.

4.7 Medium Scale Indentation Tests

The results of the Pond Inlet indentation tests conducted in 1984 were (as in many projects) confidential and proprietary for a period of time after the tests had been conducted. When I first saw the videos of the Pond Inlet indentation tests in the mid-1980s showing the extrusion of crushed ice particles (the data being confidential at that time), I had just been reading the papers by Kheisin and co-workers (Section 2.7), and it seemed that crushing with extrusion of small particles was a common feature of the two test series. This of course turned out to be true, and the observation was also found in many other interactions of ice floes with structures. Yet the identification in Kheisin's work of a distinct layer of microstructurally modified material seemed highly relevant and indicated that the same would be found in the medium scale tests, as also turned out to be the case. These observations provided the cornerstone of the approach in the present work to damage mechanics. Since the ice behind the layer is observed to be relatively undamaged, it was apparent that significant energy is absorbed in the layer itself, an indicator that the ice response had been changed due to the change in its microstructure.

The medium scale indentation tests were originally designed as support for the Hibernia project to further knowledge of the pressure–area scale effect, whereby the average pressure on global areas is smaller than pressures on smaller areas. Masterson et al. (1992) give details of the system, while Masterson (2018) provides a fascinating and entertaining account of the development of this innovative and demanding project. The first test programme was conducted by the Hibernia Joint Venture Participants (with Mobil Oil Canada as the operator), who commissioned the apparatus. The tests were conducted at Pond Inlet on the northern Baffin Island in Nunavut. The Hibernia apparatus was first deployed in tunnels excavated in a grounded iceberg, and then in a trench excavated in a thick multiyear floe in the Canadian high arctic, and finally two series in trenches in multiyear ice, part of an ice island (Hobson's Choice Ice Island, which was the site of a Canadian research station).

A second series using the medium scale apparatus was carried out during the winter of 1984–1985 in the Byam Martin channel, Rae Point (Masterson et al., 1999). The apparatus was then donated by Mobil Oil Canada to Memorial University of Newfoundland. Two series of tests at Hobson's Choice Ice Island (Frederking et al., 1990; Masterson et al., 1993) were conducted in 1989 and 1990. Preparation on site for testing required considerable effort. I shall describe very briefly the site preparation for

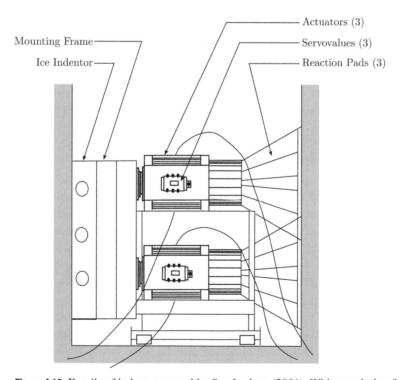

Figure 4.10 Details of indentor assembly. See Jordaan (2001). With permission from Elsevier.

Figure 4.11 Indentor assembly in trench. Photo courtesy: Bob Frederking (in the foreground).

the second of the Hobson's Choice test series. The trench was first cut using a "ditch witch" chain saw apparatus and then excavated with a Caterpillar loader with ripper bar. Figure 4.10 shows details of the indentor apparatus, while the assembled indentor apparatus is illustrated in Figure 4.11. Most of the tests were conducted with three actuators and reaction pads, but in the 1989 Hobson's Choice and in three of the 1990 series, a single actuator was used. In the case of the latter tests, this permitted a faster loading rate, up to 300 mm s^{-1}.

The original objective, to investigate the scale effect, was achieved in the first programme at Pond Inlet with global areas up to 3 m^2. One might then consider the results as input to local pressure design, as there are many results for a range of indentor areas, especially for the Pond Inlet series. In my view, the test geometry is very confined compared to real-life field interactions; one has only to compare the experience of the Molikpaq where the local areas experienced pressures only a proportion of the period of global interaction, perhaps for 10 min during an hour-long global interaction. In the case of the indentor tests, the indentor is forced into a small local area, imposing high pressures on the area in a continuous fashion over a period of seconds. In considering local pressures, it is difficult to relate the loaded area in the indentor series to a real full scale interaction such as one with the Molikpaq structure, where the width of structure approached 90 m, and we are considering a local area within that. As a result, exposure in the indentation tests is difficult to define and to compare with measurements on working structures in the ice environment. How can one compare an interaction

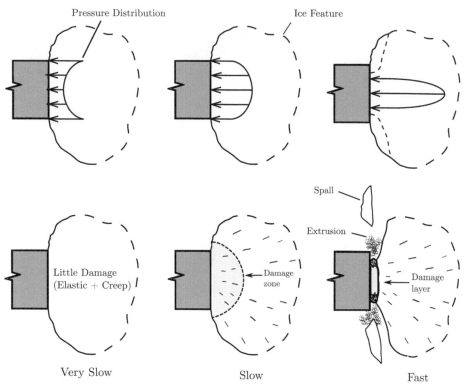

Figure 4.12 Schematic illustration of the effect of indentor speed of loading on stress distribution and on ice response (Jordaan et al., 2016). With permission from Offshore Technology Conference.

lasting seconds using an indenter with, say, an hour–long interaction with the Molik-paq where a panel is loaded only some of the time? We shall not pursue the use of the medium scale data for local pressure evaluation.

Yet the medium scale tests have provided important information regarding the failure mode of ice in compression. The reason is that the performance of the tests on site, under controlled conditions, allowed the state of the ice after the test to be carefully examined. This proved to be a significant step in uncovering the mechanism of ice failure. It was possible to look at the microstructural changes in the ice in detail. These aspects will be given attention in the following. As background to bear in mind, Figure 4.12 illustrates in summary form the behaviour of an ice specimen under indentation, in particular the development of damage (in terms of microstructural change) at various rates of loading. The changes in microstructure and development of "damaged" layers with properties distinct from that of the parent ice, becoming more important as speed increases, should be borne in mind.

In the Pond Inlet test series, tunnels of depth about 15 m were excavated into the grounded iceberg, while the physical dimensions of the apparatus dictated an average tunnel area in cross section of 3 m × 3 m. Five spherical indentors of areas varying from 0.02 up to 3 m^2 were used. The velocity followed a half-sine curve from

(a)

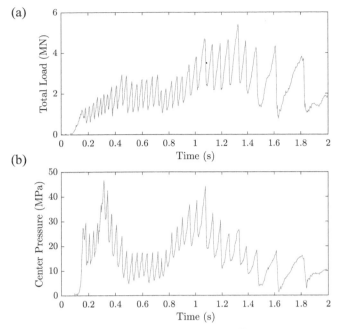

(b)

Figure 4.13 Results of Pond Inlet test T2T6, 1 m^2 indentor. (a) Total load versus time. (b) Centre pressure. See also Kennedy et al. (1994). With permission from Elsevier.

a value of 10 cm s^{-1} to zero. Pressures on small sensors (19 mm diameter) placed in the indentor surface measured local pressures within the indentor. Several results are of importance. First, the failure mode and pressures were quite unlike those that would be found in laboratory specimens uniformly and uniaxially loaded. This is in itself unsurprising since indentation and uniform loading are known to be different. Very high average pressures appeared on the smaller indentors, which declined with increasing area to values of about 3 MPa on the larger areas, supporting the idea of decreasing global pressures with area. These and other results were summarized in Masterson and Frederking (1993). The load traces showed regular sawtooth variations indicative of ice-induced fluctuations. Figure 4.13 shows results for 1 m^2 test T2T6 of the Pond Inlet series. The frequencies of vibration are also summarized in Kennedy et al. (1994). Masterson (2018) noted that "The sawtooth imposed on the force-time curve was not anticipated in 1984." In retrospect, it could be said that the velocity of interaction was in the "sweet spot" for vibrations. Very high local pressures, the highest being about 70 MPa, were measured on the small local pressure sensors (diameter of the order of 2 cm) within the indentor area.

Vertical sections were sawed in the ice and removed from the test face after the test, and were photographed. A clear distinction between the crushed layer and the parent ice was generally found. The crushed layers were digitized (see Kennedy et al., 1994 for details). Layer thicknesses averaged between 25 and 65 mm. Subsequent work in which thin sections were taken and studied showed in more detail that there were zones where the layer was not evident except in thin sections. These zones would generally

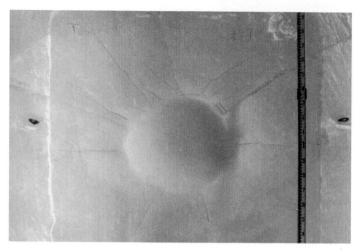

Figure 4.14 Appearance of ice surface after slow test at Rae Point, 1.0 m^2, speed = 0.3 mm s^{-1}. See also Masterson et al. (1999). With permission from ASME.

correspond to ice that had been severely recrystallized into small grains but was not yet extruding, possibly the zones where crushed ice was formed under high pressure. This will be discussed further below. Kennedy et al. (1994) proposed an analysis based on linear viscous theory (following Kheisin), resulting in a paraboloid-like bulge of pressure.

The remaining test series were conducted in trenches made in multiyear ice. The trenches were 50–90 m long, depending on the series, and a variety of indentors were used. The Rae Point series (Masterson et al., 1999) focussed on tests at different velocities, using for the most part 1.0 and 2.5 m^2 spherical indentors. The results generally confirm the trends illustrated in Figure 4.12 discussed above. Figure 4.14 shows the ice surface after a "ductile" failure mode observed at slow indentation speeds. Large ice wedge failures were often associated with the ductile failure modes. These are taken to be associated with the long duration of the slow ductile tests, and will be discussed further with regard to results from the Hobson's choice tests. The "transition" from "ductile" to "brittle" was found to occur at about 3.2 mm s^{-1}. But this is misleading, as fractures occur at all speeds, some large, time-dependent ones at lower speeds. In reality ice behaviour is more of a continuum, with fracture occurring at all speeds. There is a transition to layer formation and extrusion as speeds increase, with fractures at all speeds. The "transition" is also discussed in Section 2.9.

There were two series of tests at Hobson's Choice Ice Island: the 1989 and the 1990 series. In the first, rigid spherical and circular flat (flexible) indentors were used, both 0.8 m^2 area, as well as a rigid flat rectangular indentor 0.5 m high, 0.75 wide of area 0.375 m^2. In the second series, large and small flat-flexible indentors (0.7 and 1.8 m^2), rigid wedge indentors (1.8 m^2) and flat-rigid indentor (1.8 m^2) were used. The flexible indentors were designed to allow some bending deformation during the test. The results have been summarized in Frederking et al. (1990) and Masterson et al. (1993), respectively. An important result obtained from the 1989 Hobson's choice

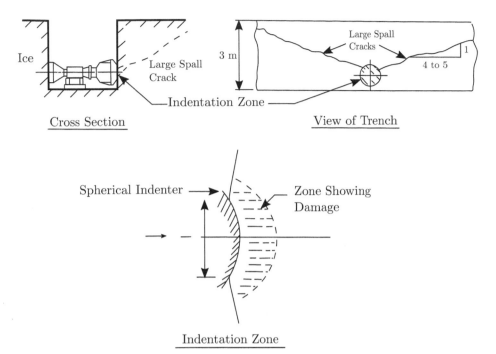

Figure 4.15 Sketch of the result of test NRC 01, showing damaged zone and large crack. See Jordaan (2001). With permission from Elsevier.

was the contrast between slow versus fast indentation, adding to the results from the Rae Point tests. We consider first the results illustrated in Figure 4.15, in which an indentation rate of about 0.3 mm s^{-1} was used. A permanent depression in the ice face, with evidence of substantial microstructural change in the vicinity of the indentor, was observed, similar to that shown in Figure 4.14. In these tests, neither ejection of crushed ice nor localized spalling were observed. However, a large spall crack in the Hobson's choice test resulted in creation of a discrete ice piece 20 m in horizontal extent as shown in Figure 4.15. The damaged zone noted was found in thin sections to be significantly recrystallized into small crystals but there was no distinct layer, with the extent of the damaged area larger than that found in the layers associated with faster loading (Frederking et al., 1990; Jordaan and Xiao, 1992). This result adds detail to the discussion above under "ductile" results for the Rae Point tests.

We describe now the response of ice to faster loading using first test NRC 05 of the Hobsons's Choice 89 series. The loading rate was about 90 mm s^{-1}, and a single *hpz* near the centre formed. Spalls and damaged ice with extensive microfracturing were in evidence in the outer areas, with a bluish interior. The thin section shows extensively recrystallized material in the blue zone (shown in Figure 13 in Jordaan, 2001). Regarding the 1989 series, Sinha and Cai (1992) state that "The so-called 'blue' zone is not necessarily undamaged ice. Severe grain modification, primarily recrystallization, is the major feature of these zones and can reach depths of more than 100 mm". This is important since some claims had been made that the ice in the blue zones was "intact".

(a)

(b)
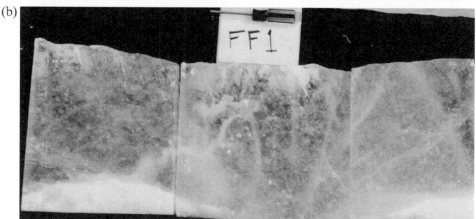

Figure 4.16 Test TFF1. (a) Test face with section removed for thick and thin sectioning. (b) Thick section from test face shown above. Photo courtesy: Bob Frederking.

During faster indentation, cyclic force–time behaviour, with layer development and extrusion of small particles, was observed, as in the earlier Pond Inlet and Rae Point tests. The layer of extruding ice has been shown in the medium scale tests to be associated with this dynamic activity. Later results from the Hobson's Choice series, which we shall now describe, show that the areas in which a layer was not apparent very likely corresponded to zones in which the ice was severely recrystallized but without the whitish colour associated with microcracked zones. These would correspond to the centre of extrusion and to the region where crushed ice is formed under high pressure prior to extrusion. Recrystallization and microfracturing in a distinct layer are also features of faster indentation. The time-dependence of fracture processes is an important factor; we note here and elsewhere (Section 9.1) the totally different fracture patterns at slow and fast loading.

We shall describe in some detail the results of a second test (TFF1) from the 1990 series, in which a flat flexible indentor acted against a wedge ice shape. The initial contact area was 300 by 1500 mm (horizontal to vertical dimension) with an ice slope at the edge of 3:1 (lateral dimension to that in direction of indentor movement). Speed was 100 mm s^{-1}. Figure 4.16 (a) shows the test face after testing with a section of the ice removed for examination and thin-sectioning, and Figure 4.16 (b) shows the section removed. Figure 4.17 shows thin sections from the centre and side of the indentor corresponding to Figure 4.16. The material in the centre layer is again highly recrystallized. This corresponds to the centre of extrusion, where the lateral deformation contributing to the extrusion is minimal. The velocity of extrusion and associated strains vary from a negligible amount at the centre of the high-pressure zone to a maximum at the edge. The layer would be expected to be less developed near the centre of extrusion, where there is a "stagnation point" and the creation of the crushed ice material. The blue colour is evident where there is no extrusion channel, and would

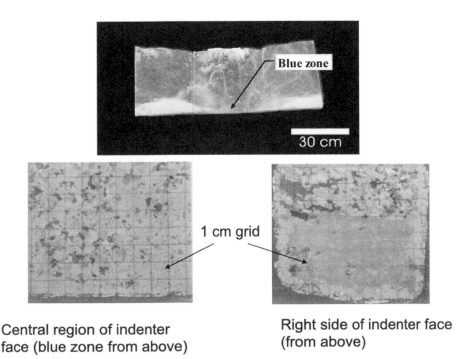

Central region of indenter face (blue zone from above)

Right side of indenter face (from above)

Figure 4.17 Thin sections from thick sections of Figure 4.16. Photo courtesy: Bob Frederking.

correspond (as noted) to the proximity of the centre of last extrusion, and near the area of highest pressure in the last cycle.

One very important conclusion from the medium scale tests is that local pressures on small areas can be quite high; values up to 70 MPa have been measured. Similar peak values had also been found in the ship rams during the Louis S. St-Laurent 1980 trials. Given that the pressure sensors were placed in predetermined locations and that there is some randomness regarding the location of the *hpzs*, it is likely that the measured maxima were less than the absolute maxima. Values up to 100 MPa have been frequently found in small scale indentation tests in which measurements have been taken across the surface of the indentor (Section 8.6). This makes it clear that, coupled with the information on the microstructural changes in the ice, the usual laboratory tests on uniformly stressed specimens are not of direct relevance to the actual failure modes. The need for triaxial testing is evident.

4.8   Ice-Induced Vibrations: Feedback between Ice and Structure

The force and pressure–time curves obtained in the medium scale tests conducted at faster rates show considerable regularity in the time-dependent fluctuations. Figure 4.13 shows typical results. It had been surmised that the regular oscillations in load were associated with activity in the microstructurally modified layer, termed "layer failure", whereas the occasional sharp changes in load were the result of spall fractures. Phase-plane diagrams of the kind used to depict a narrowband stochastic process were used in Jordaan (2001) and Jordaan and Barrette (2014) to study periods of regular oscillation. These show that the compressive ice failure process can be extremely regular; on occasion, the cycles repeat themselves almost exactly. Spectral densities also show a concentration of frequency content in a narrow range. The regularity may be interrupted by spalls; the relation between the two mechanisms has been clearly demonstrated in small scale indentation tests described in Section 8.6 using high-speed video and pressure-sensitive film. The vibration constitutes a self-sustained oscillation (Pikovsky et al., 2001) in a system comprising the moving indentor which supplies a source of energy that is dissipated within the layer of microstructurally modified material by means of viscoelastic deformation enhanced by damage processes. The cycles generally represent an interaction between the ice mass and the indentor system. If the indentor is very flexible (often the case in small-scale laboratory tests), there can be a small relative movement of the ice on the upswing in load; the flexible indentor bends back absorbing elastic energy. Upon extrusion the indentor moves rapidly forward, constituting most of the forward movement. There is essentially an energy flow into and out of the indentor and the ice (Jordaan and Timco, 1988). This is termed "alternating brittle-ductile" crushing failure, or "sawtooth" interaction.

Figure 4.18, based on Kärnä and Turunen (1990) and O'Rourke et al. (2016a), shows the variation of structural deflection in time with speeds of ice movement increasing from (a) through (d) in the figure. Vibrations in offshore structures have been observed in many situations. Prominent amongst these are slender structures

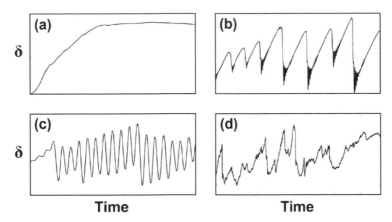

Figure 4.18 Structural deflection δ under increasing ice speed: (a) creep or damage-enhanced creep, (b) sawtooth, (c) lock-in and (d) random ice failure. See also O'Rourke et al. (2016a). With permission from Elsevier.

such as lighthouses and route markers in the Baltic Sea. In many instances, resonant behaviour of these slender structures has been observed, and has led to failure of some structures. Work by Määttänen (1983, 2001) has focused on these problems and provided initial analytical and practical solutions. Peyton (1968) and Blenkarn (1970) observed vibrations in Cook Inlet, Alaska, on offshore structures, including a test pile and jacket structures with tubular members. These were again slender structural members. Structures in the Bohai Bay have also experienced ice-induced vibration, as well as in several other regions. As noted above, Peyton believed that the cause of the "ratcheting" might be within the ice whereas Blenkarn emphasized structural flexibility. (It is our belief that both of these factors should be taken into account.) Both Peyton and Blenkarn emphasized the rôle of ice velocity with slower velocities being associated with the ratcheting effect.

Määttänen (2001, for example) developed a theoretical approach, which has considerable merit, to ice-induced vibration of the structure and to resonance. He based the analysis on the variation of peak stress with stress rate. It will be observed that there is first a positive slope and then a negative one, as the rate increases. He then developed the equations of motion and inserted this relationship. If the equation of motion, written in differential form, has positive roots, then dynamically unstable conditions arise. The approach is good, but the essentially uniaxial condition in the ice, as used in his analysis, should be extended to account for the failure of an *hpz*, which involves complex triaxial states of stress and associated mechanics, as outlined in Part II of this work.

The well-known incidents of the Molikpaq during the 1985–86 season, during which several severe ice–structure vibrations occurred, are important. Indeed, during the deployment the previous year at the Tarsuit P-45 site during 1984–85, there were many ice interaction events involving ice crushing. Rogers et al. (1986) made the following observation: "An important conclusion of the event analysis is that high load events tended to coincide with low drift speeds in the range 0.05 knots to 0.2 knots. At these speeds ice exhibited a "pulsating" mode of failure where the ice sheet seemed to act as a battering ram with a frequency of impacts ranging from 0.3 to 4 Hz. In fact it is likely that the flexibility of the structure accentuates this forced vibration process since different ice conditions result in similar pulsations." These were generally of the "sawtooth" type of oscillation. There were several interaction events during 1986 that involved dynamics, in particular the 12 April 1986 event. Liquefaction of the sand core occurred in the key events, causing considerable softening of the core. Partial evacuations of the installation occurred on 12 April and 12 May 1986. The high-pressures zones are intimately associated with the deformations of the structure, and vice versa. There is a coupling that is the root of the dynamics; this will be discussed further in Section 8.6.

4.9   Field Tests on Iceberg Failure

During the summer of 1995, full-scale iceberg impact tests were conducted on Grappling Island (near Cartwright, Labrador, Canada), by the Centre for Cold Ocean Resources Engineering (C-CORE), and other partners (Ralph et al., 2004). The aim of this field test programme was to measure the loads generated during the impact of towed icebergs against a fixed instrumented panel. Impact forces and pressures were measured on a 6 m by 6 m instrumented panel mounted on a nearly vertical cliff face. Ice masses of bergy bit and growler size were towed into the panel by an ocean-going tug. Twenty-eight impacts were recorded, 22 of which were used in the analysis of pressure–area relationships. The measured crushing pressures were typically lower than values obtained from other medium scale tests at corresponding contact areas. Localization of forces and layer development were found, as in other interactions. A sample of ice was retrieved from the impacted surface of an iceberg, and transported to St. John's (Muggeridge and Jordaan, 1999). The sample was thin-sectioned to investigate the microstructure. A clear boundary between the "parent" iceberg ice and crushed and recrystallized material was found. The penetrations were small compared to those in the medium scale tests, being limited by the mass of the ice pieces.

5   Design for Ice Loading

5.1   Background and Lessons from Mechanics

In many offshore locations where ice is present, there is a need for engineering activities and structures to be built; due account has to be taken of ice and the loads that it imparts to the structure. These activities include design of installations and of ice–strengthened vessels. Design for ice loading is undertaken for global and local pressures, as defined in Section 3.1. As noted there, the first ensures integrity of the structure as a whole, for example resistance against sliding or overturning of a gravity-based structure, or to ensure adequate hull girder strength of a ship. The second, local pressure design, deals with the resistance of plate and structure over local areas of the structure. In brief, global loads represent the maximum total force applied to the structure, while local loads represent the force (or pressure) on particular areas such as the plate between frames or other structural elements assembled. The background summarized in Section 3.1 is important to understand the approach to global and local load and pressure determination for design.

The present work focusses on design methods that have been developed alongside the research on ice failure and related fracture processes. We shall not review the history of the development of past methods to determine design parameters. We refer the reader to the books by Sanderson (1988) and Palmer and Croasdale (2013), especially the latter on design issues. Reference to the work of Croasdale (Croasdale and Marcellus, 1981; Croasdale, 1984) on limit force, limit stress and limit momentum (or kinetic energy) is most useful. This is also summarized in ISO 19906:2019, which provides good guidance. Yet in many cases, one can do a more complete and effective analysis, going deeper into the mechanics than is stated in codes of practice. I do not support "blind" use of any code; it is preferable to be able to supplement the approach suggested with deeper analysis. It is important that studies of ice mechanics recognize the time-dependency of all deformational process of ice. Not only is fracture and associated localization time-dependent (see Sections 2.1 and 2.9 and also Chapter 9), but the nature of the localization is also a function of loading rate; see Figure 4.12. As soon as a damaged zone develops, the behaviour of the ice alters, often very significantly. Localization of forces into a fraction of about 10% of the global area occurs during ice crushing. We shall consider these phenomena further in Part II of this work, but they should be understood in general terms in proceeding to design. Local pressures generally follow the (–0.7) decline with area, with global pressures declining with area approximately to the power (–0.5).

5.2   Approach

Our approach is based on probabilistic methods, consistent with ISO 19906, so that we aim at extreme-level (EL) and abnormal-level (AL) events. These are specified at annual exceedance probabilities of 10^{-2} and 10^{-4}, respectively. In brief, the code is based on a limit-state format, calibrated to obtain a target safety level of 10^{-5} per annum. Some loads (wind, waves, for example) are relatively frequent so that design events are captured within EL. Others (earthquake, iceberg impacts in some regions, for example) are rare, with long return periods, and are captured within AL. Design checks for any loading case should be made for both EL and AL events, associated with the ULS (ultimate limit state) and the ALS (abnormal limit state) respectively.

In designing, the consequence set is most important. ISO 19906:2019 includes the following statements (quoted with premission from ISO).

Clause 7.2.1.2: "The ULS requirement ensures that no significant structural damage occurs for actions with an acceptably low probability of being exceeded during the design service life of the structure."

Clause 7.2.1.5: "The ALS requirement is intended to ensure that the structure and foundation have sufficient reserve strength, displacement or energy dissipation capacity to sustain large actions and other action effects in the inelastic region without complete loss of integrity." and "The safety for each abnormal and accidental limit state shall be verified by ensuring that the effect of combined actions does not cause global failure or collapse of the structure or any of its vital parts."

The use of the methods described require an estimate of the number of events per unit time in the practical situation being analysed. This is a key activity to determination of EL and AL loads. For instance, will an icebreaker be required to carry out a few, or hundreds, or even thousands of rams into thick ice per year, as considered in Carter et al. (1992)? The report just quoted included a clear and explicit relationship of arctic class to exposure. Rather than rams, the number of interactions (glancing or direct impacts) per unit time could be used. An attractive approach is to use the upcrossing rate method (Li et al., 2010). This aspect is central to extremal analysis (Jordaan et al., 2005a), exemplified by the old fisherman's saying "the longer you fish, the bigger the fish you catch". We take the unit of time as a year, so that we can reference annual risk. In some cases, the probability per year is estimated at a value far less than unity. For instance, the probability of collision of an iceberg with an offshore floating installation on the Grand Banks of Newfoundland might be 0.01 per year, with even lower values for parts of the Barents Sea. For this probability of collision, and a return period of 10,000 years (10^{-4} annual exceedance probability), we are dealing with the maximum of 100 events. The two situations of frequent and rare load events are illustrated in Figure 5.1, from Jordaan (2005), based on the parent probability distribution of load and the arrival rate.

The use of the Poisson distribution for modelling ice impacts has been introduced in Sections 3.1 and 3.5 in describing the alpha method for local pressures. Jordaan et al. (1987) suggested an outline model for loads in arctic shipping. Ships and disconnectable floating systems can avoid collisions by manoeuvring. The key from a

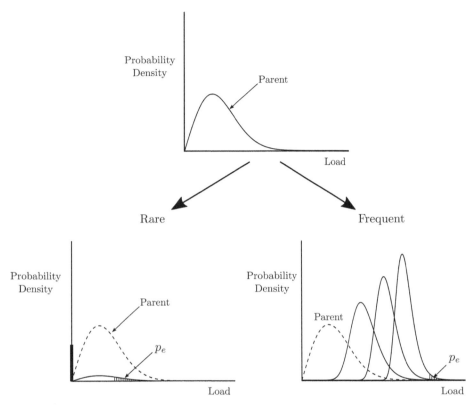

Figure 5.1 Rare and frequent loading processes. From Jordaan (2005). With permission from Cambridge University Press.

probabilistic point of view is the number of repetitions of load applications per unit time, generally taken as a year. In all cases, attention to full scale data, where available, is essential, as this provides the best guide to realistic load estimates. Freeman Ralph has led several studies in this area (Ralph et al., 2006; Ralph and Jordaan, 2013). Models for determining iceberg loads and pressures can be developed by simulating the mechanics of collision including associated probabilistic models using Monte Carlo methods. These can include factors such as mass, velocity, eccentricity of collision, and iceberg shape. C-CORE's ILS (Iceberg Load Software) is a software that achieves this. Background and theory are given for example in Stuckey et al. (2008) and Stuckey and Fuglem (2014) with an application to an arctic floater in Jordaan et al. (2014). C-CORE has also developed a Sea Ice Load Software (SILS); see, for example, Fuglem et al. (2014).

5.3   Local Ice Pressures

The results obtained from the field programmes described in Chapters 3 and 4 have provided a rich opportunity for developing rational methods of design. Local pressure data from ship impacts were discussed. For local design, an area corresponding to the panel or structural area under consideration for structural design is studied. In this case, the pressure variation over time on the local design area has been analysed, and the maximum for the event is considered. The events in the analysis of data have a certain duration, but they are all characterized by an exponential form in the extreme. The IDA project included an analysis of several data bases, and in particular the rather rich Polar Sea data base (Table 3.2). In using these analyses, the idea would be to choose those sets that correspond to the geographical area under consideration. Operational details and strategies for the new development would be taken into account in the exposure terms.

The design values for EL and AL conditions would be carried out using equation (3.20):

$$z_e = x_0 + \alpha \left\{ - \ln \left[- \ln F_Z \left(z_e \right) \right] + \ln \nu + \ln r \right\}, \tag{5.1}$$

and equation (3.21):

$$z_e \simeq \alpha \left(\ln n + \ln \mu \right). \tag{5.2}$$

In using discrete interactions with ice or ship rams, duration of impact in the design situation is compared to the duration of the interactions in the data base, and exposure adjusted accordingly. This is described in equation (3.19).

The data sets discussed are, as noted, rather rich in variety and potentially provide valuable guidance for design of future projects. The probabilistic methodology has been used in several studies and projects by C-CORE, in particular with regard to offshore resource development. Ralph et al. (2006) and Ralph and Jordaan (2013) discussed the design of arctic ships and vessels for use off the east coast of Canada. Jordaan et al. (2014) described the design of an arctic floater, for use in areas such as the Grand Banks, the Flemish Pass, and the Barents Sea. Widianto et al. (2013) described the approach to the design of the Hebron gravity-based system, and Fuglem et al. (2020) described disconnect criteria, in part using the methodology described in this section. Application to structures in the northern Caspian Sea is outlined in Jordaan et al. (2011).

The approach has also been applied in assessments of design ice loads for a transit operation along the Northern Sea Route in Ehlers et al. (2017). This provides a mission-based design methodology. It has also been applied in the analysis of loads measured in the field for the reduction in local pressure using ice management (Fenz et al., 2018). There are many opportunities to match the required conditions with the variety of field conditions in the data set, and the methodology allows the use of any arbitrary exposure.

Clause A.8.2.5.3.1 in ISO 19906:2019 deals with short duration impacts with fixed structures and floating systems, and follows the reasoning outlined above. Clause A.8.2.5.4 on local pressures for thick, massive ice features is based on an extremely conservative analysis using results from medium scale indentation tests, flat jack tests (a very unsuitable method for local load estimation), and on maxima from unadjusted Medof panel readings. The result is a highly conservative set of design values, which also leads to pressures on larger areas equal to 1.5 MPa, higher than any reliably measured value. It is as well to conduct a detailed analysis of *hpz*s, their areal density and strength to determine design values, including background pressure accompanying the local design values. Background pressure is best treated as a load combination issue and does not pose a significant problem unless unrealistic values (such as the 1.5 MPa value) are used.

5.4   Global Ice Pressures

In design, generally the most demanding loading condition comes from an ice crushing interaction, which occurs certainly in icebreaking and is also prevalent in most offshore regions. The exception might be areas where landfast ice moves slowly under the action of wind; there is also the possibility of slowly moving ice further offshore, as experienced by the Molikpaq (Event f603251302 in Rogers et al., 1998; see also Hardy et al., 1996). In these cases, a nonlinear creep analysis may be performed (e.g. Ponter et al., 1983). This should recognize the role of creep enhancement associated with damage processes. In all cases we need to know the global force associated with the global interaction area. Considering now ice crushing events, the area of the *hpz*s will, as noted, be a small proportion of the global interaction area. It is not possible to solve the problem using ice mechanics, and an empirical approach is the best way forward. The global areas are generally calculated from the interaction mechanics and a pressure–area relationship. For an iceberg collision, we would consider the mass, shape, velocity, eccentricity, together with the global pressure-area relationship, to calculate the global force and area.

There have been several analyses of global pressure-area relationships. Models for global average pressure have generally been formulated using power-law formulae (see for example Sanderson, 1988); note also equation (3.1). We shall discuss the following form

$$p = C(\text{global area})^D = Ca^D, \qquad (5.3)$$

where p is the average global pressure. In the original formulation, C and D were constants. We can normalize the area by dividing by a reference area, say equal to 1 m^2, then area a is dimensionless, and C has the units of MPa. Sanderson (1988) includes discussion of data for such a curve; an approximate average curve used by the present writer is given by $p = 3a^{-0.4}$ in MPa.

Clause A.8.2.4 in ISO 19906 contains a good summary of global ice actions and the various influencing factors. The clause is a rather conservative statement, but useful in formulating an approach. Ice crushing is given appropriate attention. It is noted that data derived from the Molikpaq experience has been used, and reference is made to Section 4.4 of the present work and Figure 4.7 (a), and the associated arguments related to the reliability of the Medof panels; use of the extensometer readings for global pressures is advocated. The discussion at the end of Section 4.4 on exposure is also emphasized. The fact that during ice crushing the loaded area comprising *hpz*s is of the order of 10% of the global area should be appreciated. Both fixed structure–ice sheet and ship ram based data are given attention in the ISO document.

The average values of pressure in *hpz*s are of the order of 0.5 MPa, to which must be added the contribution of the standard deviation of the averaged *hpz* values to obtain global pressure, a small addition. This results in values less than 1 MPa. As noted in Section 5.3, the values proposed in Clause A.8.2.5.4 on local pressures for thick, massive ice features in ISO 19906:2019 are rather conservative, with a limit for larger areas approaching 1.5 MPa. The same comment is made with regard to background pressure, accompanying a local pressure analysis; values of this magnitude seem unrealistic. Field measurements—such as the STRICE and Molikpaq data sets—can be used to formulate design global pressures for structures in specific locations. Appropriate analysis of exposure for the location must be conducted, and of course, the statements made regarding the Molikpaq data should be taken into account. STRICE data were used to design structures in the northern Caspian Sea, as outlined in Jordaan et al. (2011b).

Before discussing past global pressure–area analysis for vessels, it is useful to review the question from the perspective of the local pressure *hpz* analysis. Ice failure in compression is through high-pressure zones at all rates of interaction except very slow creep-type failures. There may be a small amount of force involved in extrusion of crushed ice products, but the main source of global load lies in the *hpz*s themselves. The sum of all *hpz* forces should then provide a good estimate of the global force. As an example, we assume that a ship ram results in a bow force of 28 MN with a global area of 40 m^2; these are typical values for the MV Arctic or Oden. The total number of *hpz*s is given by 28 MN divided by the mean zonal force, say 0.5 MN, giving about 56 *hpz*s, which we multiply by the mean zonal area, say 0.08 m^2, resulting in an area of 4.5 m^2, representing the total area of *hpz*s alone. This is about 11% of the global area, as expected, a small percentage.

The discussion just made regarding the sum of *hpz* forces being equal to the global force leads to an approach for design. The *hpz* forces could be modelled as being point loads random in size; an exponential distribution works quite well; see Section 3.4 and equation (3.8). These could be treated as being independent and identically distributed, with the areal density of *hpz*s being modelled as a Poisson process as in equation (3.11). Monte Carlo methods can then be used to obtain global pressures and also background pressures in a probabilistic analysis. The statistics can be developed for any of the IDA sets of rams and interactions, giving input parameters on *hpz* magnitude and areal density. Checks have been made regarding the assumption of independence and the conclusion is that the approach just suggested is a reasonable way forward.

One of the later innovations was to treat C and D in equation (5.3) as being random, as in Carter et al. (1996). In determining distributions for these parameters, one important aspect was the determination of the global area. Local pressure analysis is described in detail in Section 5.3; the area under consideration is defined by the structure. The data for a particular area could be ranked and plotted as illustrated in Figure 3.4. For global loads, the area is different for each interaction, and is not measured, and estimation varies depending on the details of the ice failure process, spalling

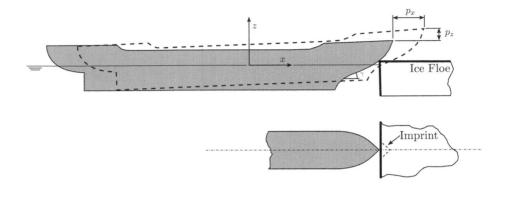

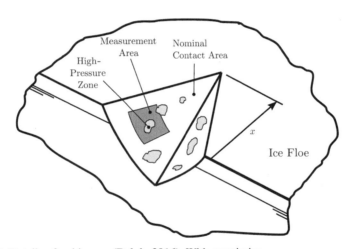

Figure 5.2 Details of a ship ram (Ralph, 2016). With permission.

and related mechanics. To overcome this, a novel method was used. This consisted of a trial-and-error approach in which assumptions were made regarding the values of C and D in equation (5.3) and comparisons to the measured values were made. An optimization approach was used to obtain the final values for the distributions of C and D and the parameters of the distributions. An important component in the analysis is that the interaction of the ship and ice involves movements of the ship in response to the force of interaction. Figures 5.2 and 5.3 illustrate the process and phases of ship ramming. The "Fmax" computer program and associated analysis were formulated to deal with this problem. The program was developed by Mark Fuglem and described first in Carter et al. (1996) as well as in IDA (2007) and Li (2007).

The two-dimensional ramming model (Fmax) accounted for surge and heave. Stiffness in pitch was modelled by an equivalent spring stiffness at the bow. Vessel and added mass were modelled by equivalent masses at the bow together with damping

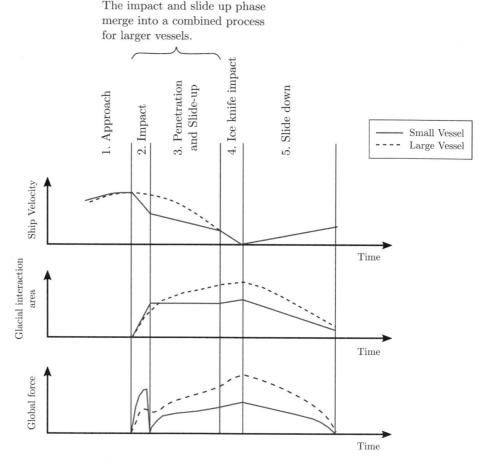

Figure 5.3 Phases of a ship ram (Ralph, 2016). With permission.

coefficients. The ship was modelled with a single stem and opening angle. The ice was modelled as a plate, with the area of contact governed by the pressure–area equation (5.3) with particular chosen values $C = c$ and $D = d$, for example $C = 3.0$ MPa, $D = -0.4$. Other aspects of the mechanics of interaction were also considered: friction at the bow-ice interface, flexure of the vessel, inclusion of a skeg, flexural failure of the ice, and floe tipping. The equations of motion were solved in time-domain using the Runge-Kutta method. The crushed ice was modelled including a "soft" layer. A trial-and-error approach was followed, using Monte Carlo methods and varying values of C and D. Vessels used were the Kigoriak, the MV Arctic and the Manhattan, with checks using results from the USCGC Polar Sea and the Oden.

The main result was presented in Carter et al. (1996) and Jordaan et al. (1996), and is given in ISO Clause A.8.2.4.3.5 (Global ice pressures from ship ramming tests). The quantities C and D were modelled using lognormal and normal distributions, respectively, and then a^D is also lognormally distributed. Consequently p is also lognormally distributed. The best result was found with $\mu_C = 3$ MPa, $\mu_D = -0.4$, $\sigma_C = 1.5$ MPa, $\sigma_D = 0.2$, where μ = mean, σ = standard deviation. In practice, the model gives relatively good results for small numbers of interactions but overestimates significantly the loads for large numbers of interactions. The lognormal distribution has a very "fat" tail, and we have noted that the global pressure, being the sum of many *hpz*s, should follow the normal distribution. Improved solutions were studied in C-CORE (2017). Fuglem et al. (2016) studied and reviewed global forces for iceberg impact loads; the results are mainly based on multiyear ice and therefore have wider applicability in terms of ice type.

5.5   The *Titanic*

Working in the area of design for structures in the ice environment, one might wish to understand, from an engineering point of view, what happened to the *Titanic*. This certainly does assist in obtaining confidence in methods that have been developed in this chapter. The RMS *Titanic* sank in the North Atlantic Ocean on 15 April 1912 after colliding with an iceberg on her maiden voyage from Southampton to New York City. The Titanic was the largest ship afloat at the time. The incident resulted in the deaths of over 1500 people, one of the deadliest peacetime maritime disasters in history. She departed Southampton on 10 April 1912, encountering calm seas and then increased speed to 22 knots. On 14 April there were reports of icebergs and at 11:40 p.m. an iceberg was spotted dead ahead. The command "hard-a-starboard" resulted in the vessel veering to port and as a result the collision with the iceberg occurred on starboard side of bow. The *Titanic* could stay afloat if any two compartments or the first four became flooded. Shortly after the collision it was found that the first six compartments were flooded. On 15 April 1912, at 2:17 a.m., the captain announced "every man for himself", and the vessel sank.

   Jordaan and Gosine (2012) provide a discussion of global and local pressures. An important point, and an obvious one, should be made at the outset. The *Titanic* was not designed for an impact with an iceberg, and this should have been recognized in its operation. Riveted connections were used in joining structural components, including the overlapping plates of the hull. The rivets were made of steel in the central section of the hull, and cast iron in the bow and stern regions. The major loading expected was associated with waves and bending of the hull, so that this strategy seems even now to be sensible. With regard to the sinking of the vessel, it seems facile to "blame" the rivets, e.g. statements such as "the fatal flaw", etc. The real flaw was in the operation, taking the vessel into an environment for which she was not designed. One first important piece of evidence results from accounts of those on board (see Jordaan and Gosine, 2012 for details), in which there was little evidence of global response of the vessel as a result of the impact. Assertions such as "no shock or jar", "it just gave a little vibration" and "there was a slight jar followed by the grinding—a slight bumping" were made. These comments suggest that the global load on the vessel was quite small, indeed negligible.

But this does not mean that intense local loading of the ship's hull did not take place. If this force is small enough in global terms, then it will not be felt significantly by those on board. The real culprit is the localization of ice loading into *hpz*s, the nature of which has been described throughout the present work. Edward Wilding, chief naval architect for Harland and Wolff, builders of the Olympic and the Titanic, testified as an expert witness at the British Inquiry concerning the loss of the RMS *Titanic*. Wilding estimated that the iceberg produced an equivalent opening along the starboard side of 12 square feet, not the 150 foot gash proposed by some. This analysis supports the observation of a series of narrow slits found in later ultrasound measurements. The discussion in McCarty and Foecke (2008) regarding slow loading rates and the fact that only a small impact was felt does not take into account the localization of ice loading, as discussed in Jordaan and Gosine (2012). In fact, the localization of loading explains the slight vibration felt during the impact but does not imply that the loading was slow. It was concluded that forces of several meganewtons, possibly as high as 7 MN, would be generated on small areas of the order of a square metre, and the force would be localized within that area.

A review of literature showed that the strength of the rivets was less than 0.2 MN (20 tonnes). The high local loading from the ice impact, while not anticipated in design, could lead to failure. The failure would be accentuated by the fact that once some rivets had failed, the opening behaves as a crack, with stresses at the end proportional to a function of the length of the opening. Rivets were somewhat greater than an inch in diameter, with spacing of about 4 diameters (Woytowich, 2003). It seems evident that an *hpz* force would be quite sufficient to cause the rivets to fail. To conclude, returning to the question of the small impact felt by passengers: this can be explained by a comparison of the force associated with high-pressure zones, say, 10 MN at most, with the displacement of the vessel, 52,000 tonnes. A simple calculation shows that the resulting acceleration is small.

Over time, riveted construction has been replaced by welded joints. Attention must be paid to the quality of the steel construction and associated welding to ensure that brittle fracture does not occur.

5.6   Codes and Standards

We mention briefly codes of practice that are relevant to design for structures and vessels in ice-prone waters. The appropriate standard for offshore structures is ISO 19906:2019. This is a good limit-states code, with excellent coverage of probabilistic methods and exposure to ice conditions. It has been referenced and discussed in Sections 5.1–5.4. Earlier the Canadian standard CSA S471 "General requirements, design criteria, the environment, and loads" developed the limit-states approach in offshore engineering.

With regard to design of ships, a probabilistic framework was proposed in Carter et al. (1992, 1996) for the then-revisions to the Arctic Shipping Pollution Prevention Regulations (ASPPR). The measure of exposure was based on Canadian experience in operations, and was taken as the number of ship rams per year in multiyear ice. This could be generalized to interactions, direct or glancing, multiyear ice being the principal hazard in the arctic. The revised ASPPR has been superseded by the Unified Requirements for Polar Ships developed by the International Association of Classification Societies (IACS, 2006, 2011). The IACS has established seven different Polar Class notations, ranging from PC 1 (highest) to PC 7 (lowest), with each level corresponding to operational capability and strength of the vessel. The Code relies on the extensive experience of the authors. It would benefit from probabilistic background and consideration of exposure from a risk perspective. A detailed review has been conducted by Ralph (2016), including considerations of exposure. Kim and Amdahl (2016) provide a good description of mathematical basis of the IACS and the related Russian rules (RMRS, 2014). They describe the deterministic ice edge analysis based on viscous layer theory in the Russian rules. See also Popov et al. (1967). This layer theory was introduced in Section 2.7 and is further elaborated in Section 6.4. While the layer theory constituted a breakthrough in ice mechanics at the time, the extension to the mechanics of ice–ship interaction would benefit from consideration of *hpz*s and the large set of available field measurements described in Chapters 3, 4 and the present chapter.

Part II

Theory of Time-Dependent Deformation and Associated Mechanics

Part II

Theory of Time-Dependent
Deformation and Fracture
Mechanics

6 Viscoelastic Theory and Ice Behaviour

6.1   Introductory Comments

Ice response to imposed stress invariably results in time-dependent movements: conversely, the response to imposed deformation leads to time-dependent stress changes. Understanding and analysing these effects is an imperative in ice mechanics. The main focus is compression, but when fracture is considered, tensile stresses are very important. For slow loading under compression, the deformation is mainly creep, possibly enhanced by damage. Some very large time-dependent fractures have also been observed under slow loading. As the rate of loading increases, fractures become more frequent, smaller and more localized (see also Figure 4.12). The load becomes concentrated into high-pressure zones, through which most of the compressive load is transmitted. All of the response to stress or strain is time-dependent, including the microstructural change that affects behaviour. It is important to bear in mind the high pressures and triaxial states of stress that must exist within the high-pressure zones.

We have also noted that time-independent theories that do recognize irreversible nonlinear behaviour, for example plasticity theory, do not provide adequate solutions to what is a problem necessitating time-dependent approaches. This is true even if the theory is adapted to reflect the effect of concurrent hydrostatic pressure through a pressure-dependent yield criterion. Viscoelastic models, in contrast to rate-independent plasticity-based models, exhibit deformation that continues with time under the influence of applied load. The word "plasticity" here is used in the engineering sense, that is time-independent strain with a distinct yield stress. For metals and alloys, viscoplasticity represents the macroscopic behaviour caused by the movement of dislocations in grains, as well as inter-crystalline gliding. If these movements are impeded, this results in a yield stress, above which flow occurs. Physicists often refer to the time-dependent flow as "plasticity", a practice that we will not adopt. But it is important to note that some writers use this term without implying time-independence.

To illustrate the problem in deriving values for a yield stress, Figure 6.1 shows the stress–strain response of ice (essentially the Sinha equation) under different strain rates. Two entirely different yield plateaux result, whereas plasticity theory has only one time–invariant one. This inadequacy of plasticity theory becomes much more evident when damage is included, with a large variety of creep rates covering many

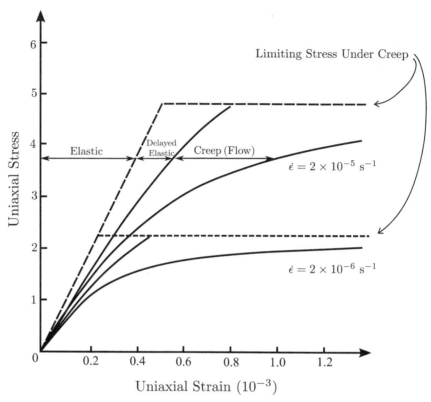

Figure 6.1 Viscoelastic analysis shows different yield criteria for different strain rates. See also Jordaan (2015). Reproduced with permission from ASME.

decades in value. There does not seem to be any basis for proposing plastic yield criteria. In Jordaan (2015), the total inappropriateness of models based on crushable foams in ice mechanics is exposed.

The answer to the dilemma is, we believe, viscoelastic theory, together with damage mechanics to reflect the changes to material response resulting from the evolution of microstructural change in time. A very attractive and basic approach to viscoelasticity has been pioneered by Biot (1954, 1958) and Schapery (1964, 1968) in which the stress–strain relationships are derived using a fundamental thermodynamic approach (thermodynamics of irreversible processes, TIP) yet arrive at practical solutions that are of use in engineering problems. This approach is summarized in Appendix A. An excellent first introduction to linear viscoelasticity has been given by Flügge (1967); other textbooks include those by Christensen (1971), Haddad (1995) and Lakes (2009). An interesting description of the mathematical structure of the linear theories is given by Gross (1968). With regard to ice, we note that Pounder (1965) got the direction of analysis correctly by emphasizing viscoelastic models, including time-dependency, rather than time-independent ones. But the high-stress triaxial regimes that occur in *hpz*s are very different from the uniaxial low-stress conditions that are usually used in determining creep functions. We shall immediately deal with the low-stress uniaxial regimes, and then progress to the more complex issues in Chapters 7–9.

6.2 Tension and Compression

Materials will generally have a different response to tension and compression. We have indicated that ice under tension is highly prone to fracture, far more so than in compression. At very low strain rates or stress, the stress–strain behaviour tends to be similar (Schulson, 1999) but the behaviour is different otherwise mainly due to fracture. There is some dilemma as to sign convention, as we deal with both compressive failure and ice failure in tension. We shall introduce basic theory with the usual convention of tension and tensile strains being positive, but will treat compression and compressive strains as being positive when dealing with damage processes. Any change in convention will be stated so that confusion will hopefully be avoided. Use of the "sign" notation, also written as "sgn", will be avoided where possible.

6.3 Low Stress Regime of Springs and Dashpots: Maxwell Unit

Ice deformation can be modelled using linear viscoelasticity under low stress levels for short periods after the application of load. The basic relationships can be used in tension or compression. Linear viscoelasticity is an important building block in the study of nonlinear viscoelasticity. We start with the spring and dashpot in series as illustrated in Figure 6.2 (a), the Maxwell model. The unit has a linear strain response to stress, modelled by a spring. At the same time, it has a viscous response, modelled by a dashpot. The viscous element models dissipation of energy. The dashpot can be thought of as a piston in a cylinder with no trapped air inside (e.g. with an aperture at the end) and lubricated to ensure steady deformation at constant stress. Furthermore, the unit has a linear relationship between applied stress σ and strain rate, constituting Newtonian viscosity. The elements have the following relationships between stress σ, strain ϵ and strain rate $\dot{\epsilon}$:

$$\sigma = E\epsilon, \tag{6.1}$$

and

$$\sigma = \mu\dot{\epsilon}, \tag{6.2}$$

where E is the elastic modulus and μ the viscosity, in MPa and MPa s, respectively.

Consider now the step increase in stress from 0 to σ at time $t = 0$, as illustrated in Figure 6.2 (b). An immediate elastic strain ensues, and stress is transmitted to the

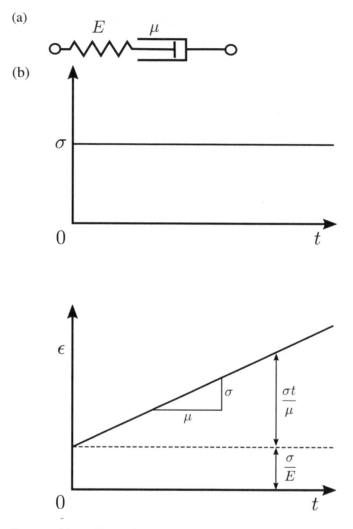

Figure 6.2 Maxwell model (a) and strain response to stress input (b).

dashpot, which begins to deform in time at the rate given in equation (6.2). Since stress is constant, the strain at time t is

$$\epsilon = \sigma \left(\frac{1}{E} + \frac{t}{\mu} \right),$$

(6.3)

the sum of the elastic and the viscous strain (Figure 6.2 (b) shows the response). This is our first example of time entering a material model, most important for the analyses to follow. It is seen that there may be a large departure from elastic conditions depending on the ratio of the viscosity μ to the elastic modulus E. This ratio (μ/E), termed the "relaxation time", gives the time in seconds for the viscous strain to equal the elastic strain for any stress σ. The creep curve for the Maxwell model is linear in time and none of the viscous strain is recovered upon removal of the load. Figure 6.3 shows the

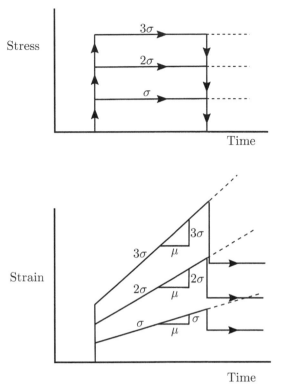

Figure 6.3 Maxwell unit response for various stress inputs.

effect of three stress levels, $\sigma, 2\sigma$ and 3σ: the strains at a given time are also linear, but with stress. This is our first example of irrecoverable strain. As noted above, elasticity implies storage of energy, viscous flow implies dissipation and heat generation.

Given an applied stress of σ at time t, the change in strain during a small interval dt is the sum of the elastic and viscous components:

$$\dot{\epsilon} = \frac{\dot{\sigma}}{E} + \frac{\sigma}{\mu}. \tag{6.4}$$

In the case where the stress σ is applied instantaneously at time $t = 0$, and then held constant ($\dot{\sigma} = 0$),

$$\epsilon = \frac{\sigma}{\mu}t + C, \tag{6.5}$$

where C can be solved by integrating over the singularity at $t = 0$. Thus

$$C = \int_{0-}^{0+} \dot{\epsilon} d\tau = \int_{0-}^{0+} \frac{\dot{\sigma}}{E} d\tau + \int_{0-}^{0+} \frac{\sigma}{\mu} d\tau = \frac{\sigma}{E}. \tag{6.6}$$

The use of Heaviside and Dirac functions assists in further analysis. They provide a convenient way of considering "jumps" in functions. The unit step function $H(t)$ is defined by the two equations:

(a)

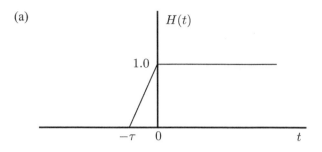

(b)

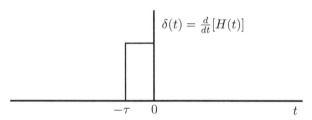

Figure 6.4 (a) Unit step function ($\tau \rightarrow 0$). (b) Dirac delta function ($\tau \rightarrow 0$).

$$H(x) = 0 \text{ for } x < 0 \qquad (6.7\text{a})$$
$$H(x) = 1 \text{ for } x \geq 0. \qquad (6.7\text{b})$$

This function is illustrated in Figure 6.4 (a), in which a small interval τ is shown to the left of the origin $t = 0$. Equations (6.7a) and (6.7b) can be conceived as a limit of the (a) function of Figure 6.4 as $\tau \rightarrow 0$. The derivative of the unit step function is a highly singular function called the Dirac delta function, $\delta(t)$, defined by the following conditions:

$$\delta(t) = 0 \text{ for } t \neq 0, \qquad (6.8\text{a})$$
$$\delta(t) = \infty \text{ for } t = 0, \text{ and} \qquad (6.8\text{b})$$
$$\int_{-\infty}^{\infty} \delta(t)dt = \int_{0^-}^{0^+} \delta(t)dt = 1. \qquad (6.8\text{c})$$

The function is illustrated in Figure 6.4 (b); in the limit as $\tau \rightarrow 0$ the Dirac function becomes a "spike" of mass unity at the origin. (As pointed out by Butkov (1968), a more rigorous analysis relies upon limiting sequences.) The unit step and Dirac functions can be shifted; $H(t - a)$ represents a unit step function "jumping" from zero to 1 at $t = a$. Similarly $\delta(t - a)$ represents a spike of mass unity at $t = a$.
    Note that

$$\int_{0}^{t} H(\tau)d\tau = tH(t). \qquad (6.9)$$

This is the ramp function. Also

$$\int_0^t f(\tau)\,\delta(\tau-a)d\tau = f(a). \tag{6.10}$$

Equation (6.4) can now be written as

$$\dot{\epsilon} = \sigma\left[\frac{\delta(t)}{E} + \frac{H(t)}{\mu}\right], \tag{6.11}$$

and integrated directly to give

$$\epsilon = \sigma H(t)\left[\frac{1}{E} + \frac{t}{\mu}\right]. \tag{6.12}$$

Ice deformation in Sinha's equation (2.4) includes a flow term

$$\dot{\epsilon} = \dot{\epsilon}_0\left(\frac{\sigma}{\sigma_1}\right)^n \propto \sigma^n, \tag{6.13}$$

where the constant $n \simeq 3$ (Glen's law) illustrates the nonlinearity with a stress-dependent viscosity $\propto \sigma^{n-1}$.

6.4   Ice Crushed Layer as Viscous Fluid

The Maxwell unit models what is essentially a viscous fluid with some elastic response, with no limit to the deformation. If we remove the spring, we are left with a dashpot that represents a Newtonian fluid. Kurdyumov and Kheisin (1976) used this model to analyse the layer discussed in Section 2.7. They applied this in a treatment of their dropped ball tests. The important new development in this work was the recognition of distinct mechanical properties (viscosity) in the layer, as against that of the parent ice. The layer is treated as a Newtonian fluid, in which shear stress $S = \mu\dot{\epsilon}$, where μ = dynamic viscosity, and $\dot{\epsilon}$ is the shear strain rate. Kurdyumov and Kheisin (1976) analysed the extrusion for their dropped-ball tests, i.e. extrusion around a spherical surface. There are analogies to lubrication theory, in which the layer thickness is taken as being very small. A variety of possible extrusion scenarios exist for different geometries of interaction, for instance extrusion in a plane from a plate acting against the structure. Figure 6.5 shows this geometry and the extruding layer.

We shall illustrate the approach by summarizing the solution of the Navier-Stokes equations for steady laminar flow between parallel flat plates. This is intended to illustrate the substance of the method with a relatively simple geometry. We use the coordinate system of Figure 6.5 but consider the flow to be between two long, flat, parallel plates under constant pressure gradient in the x direction. Flow is parallel to the boundaries, in layers, and the velocity in the x-direction, v_x, is constant, while $v_y = 0$. Pressure p is taken as positive. For equilibrium in the x-direction

$$\frac{\partial\tau}{\partial y} - \frac{\partial p}{\partial x} = 0. \tag{6.14}$$

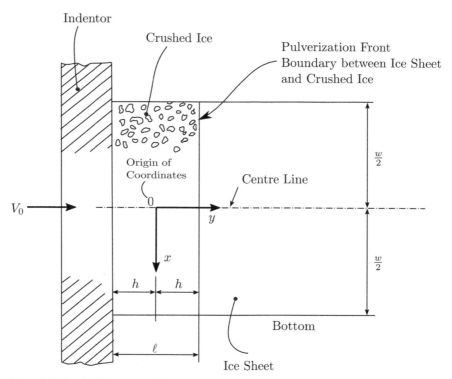

Figure 6.5 Crushed layer extrusion: geometry of system and stress state. See also Jordaan and McKenna (1988). With permission from Cambridge University Press.

But the viscosity μ enters the analysis through

$$\tau = \mu \frac{\partial v_x}{\partial y},$$

(6.15)

and then

$$\frac{\partial p}{\partial x} = \mu \frac{\partial^2 v_x}{\partial y^2}.$$

(6.16)

For simplicity of illustration, we have assumed no sliding at edges of layer; one might consider the moment before sliding, layer has been developed, extrusion is imminent but not started. (Note different boundary condition in Russian work: liquid layer on surface of solid, tangential stresses = 0, with the other condition being that the absolute velocity of particle motion near rupture surface must be zero.) We may now integrate equation (6.16) twice and solve with the boundary conditions that the velocity is zero at $y = \pm h$, to give

$$v_x = \frac{1}{2\mu} \frac{\partial p}{\partial x} \left(y^2 - h^2 \right),$$

(6.17)

where h is half the distance between the plates.

   Equation (6.17) is an exact solution of the Navier-Stokes equations for the simplified flow described. In the work of Kurdyumov and others, the equations of lubrication theory, in which the layer thickness is taken as being very small, were also used.

The viscous layer was also studied in Jordaan and Timco (1988) using the methods described. The distribution of mean pressure over a layer was found for an ice sheet failing against a structure (Figure 6.5) to be

$$p(x) = \frac{6\mu v_0}{\ell^3} \left(\frac{w^2}{4} - x^2 \right),$$

(6.18)

where ℓ is the layer thickness. A quadratic function is found, peaking at the centre, as would be expected.

The method of Kheisin and Cherepanov (1970) and Kurdyumov and Kheisin (1976) was extended in the Russian literature to ship design (Popov et al., 1967). This has been well described in Kim and Amdahl (2016). The present writer contends that the approach represented in the analysis of the crushed layer represents a large step forward in ice mechanics, insofar as it represented the first recognition of the distinct properties of the layer, as compared to the parent ice. We know from the measurements outlined in Sections 3.3–3.5 that the load in ship–ice interaction is concentrated into high-pressure zones. The loaded "patch" includes these zones. Further development must recognize the development of high-pressure zones, and the present writer has taken a different approach based on full scale measurement of ice pressures, as described in Chapter 3. Furthermore, the assumption of linear viscosity must be regarded as a first approximation as we shall see in Chapter 8.

Other works treating the extruding material as being linearly viscous have been presented (as noted) in Jordaan and Timco (1988) for sheet indentation and in Kennedy et al. (1994) for the medium scale test geometry. In these works, we do not consider the method as being appropriate for calculating extreme design loads, but the methods were used to obtain relationships for the upswing in loads during cyclic crushing, and to explore constitutive relationships. We shall see that the viscosity is highly nonlinear in a detailed analysis (Chapter 8) but the recognition of a distinct damaged layer with viscous properties was a big step forward conceptually.

6.5 Kelvin and Burgers Models

We consider first the linear case. A Kelvin (or Voigt) material has a spring and dashpot in parallel as shown in Figure 6.6 (a). One should think of rigid joints (no rotations) at the nodes of the network, with the same strain in the spring and dashpot. Applying a step stress σ at $t = 0$,

$$\sigma H(t) = E_k \epsilon + \mu_k \dot{\epsilon}.$$

(6.19)

This may be solved using Laplace transforms. Indicating the transform by the overbar, and transform parameter by s, the transform of equation (6.19) is

$$\frac{\overline{\sigma}}{s} = E_k \overline{\epsilon} + \mu_k \overline{\epsilon} s.$$

(6.20)

(a)

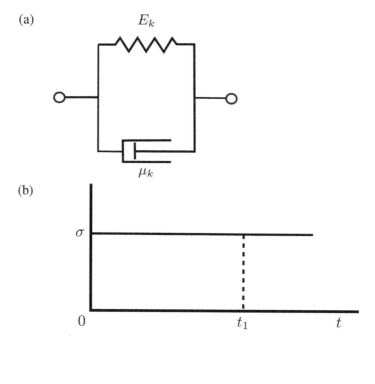

(b)

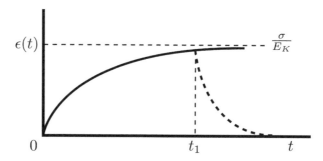

Figure 6.6 (a) Kelvin unit. (b) Strain resulting from step input σ at time 0 and its removal at time t_1.

The inverse transform of this equation provides the solution:

$$\epsilon(t) = \frac{\sigma}{E_k}\left[1 - \exp\left(-\frac{E_k}{\mu_k}t\right)\right] \tag{6.21a}$$

$$= \frac{\sigma}{E_k}\left[1 - \exp\left(-\frac{t}{t_R}\right)\right] \text{ for } t \geq 0, \tag{6.21b}$$

where μ_k/E_k is the retardation time, t_R. This is the time that would elapse for the viscous strain to equal the elastic strain with both units under the same stress. Note that we could have added the term $H(t)$ to these equations, or specify $t \geq 0$, as we have done. The equation is a representation of "delayed elasticity"; see Figure 6.6 (b).

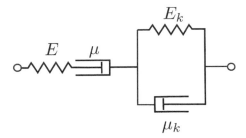

Figure 6.7 Burgers model.

The strain is recoverable upon removal of stress. This does not mean that the movement is reversible in the thermodynamic sense since the dashpot produces irreversible movements and heat generation.

The reader might wish to verify the following equations for the three-parameter solid (a Kelvin unit with a spring attached) and for the Burgers body (Figure 6.7) with stress σ applied at time $t = 0$:

$$\epsilon(t) = \frac{\sigma}{E} + \frac{\sigma}{E_k}\left[1 - \exp\left(-\frac{t}{t_R}\right)\right] \tag{6.22}$$

$$\epsilon(t) = \frac{\sigma}{E} + \frac{t}{\mu} + \frac{\sigma}{E_k}\left[1 - \exp\left(-\frac{t}{t_R}\right)\right]. \tag{6.23}$$

The Burgers model represents well the nature of ice deformation at low stresses. Kavanagh (2018) conducted fits to the low-stress tensile creep tests of LeClair et al. (1999) using linear and nonlinear viscoelastic models. Successful results were obtained in the linear case using the theory described for the short time periods considered, as shown in Figure 6.8. Nonlinear theory improved the results, but both fits were satisfactory.

6.6 Hereditary Integrals

We consider now a material with linear creep compliance equal to $J(t)$, defined as the strain corresponding to a unit stress applied at $t = 0$. The material also has a relaxation modulus $R(t)$, or stress corresponding to a unit strain applied at $t = 0$. These can be derived from each other, as we shall see in the following. The functions can be shifted. If a stress σ_0 were applied at $t = \tau$, the resulting strain at time t would be

$$\sigma_0 J(t - \tau). \tag{6.24}$$

Similarly, for a strain ϵ_0 applied at time $t = \tau$, the stress at time t is given by

$$\epsilon_0 R(t - \tau). \tag{6.25}$$

We now deal with Boltzmann superposition. Linear superposition is well known in elastic problems but we now superimpose in time. A stress σ_0 is applied at time $t = 0$, resulting in strain $\sigma_0 J(t)$ at time t. At time τ, an additional stress $\Delta\sigma$ is superimposed

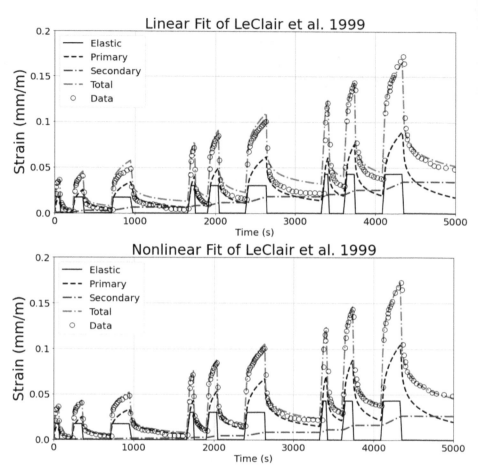

Figure 6.8 Fits by Kavanagh (2018) of linear and nonlinear viscoelastic formulations to data of LeClair et al. (1999). With permission.

on the already stressed specimen. The principle in Boltzmann superposition is that the strains are additive at times $t > \tau$:

$$\epsilon\,(t) = \sigma_0 J\,(t) + \Delta\sigma\,(\tau)\,J\,(t - \tau), \tag{6.26}$$

thus superimposing the separate strains due to the stresses σ_0 and $\Delta\sigma$, taking into account the time of application of the stresses. We may continue superimposing steps such as the one just discussed. The result is

$$\epsilon\,(t) = \sigma_0 J\,(t) + \sum\nolimits_{i=1}^{n} \Delta\sigma_i(\tau_i)J(t - \tau_i), \tag{6.27}$$

for n step loads at $\tau_i, i = 1, 2, \ldots, n$. In the limit of many infinitesmal step changes in load in the usual manner, this equation becomes

$$\epsilon\,(t) = \sigma_0 J\,(t) + \int_0^t J\,(t - \tau)\,\frac{d\sigma\,(\tau)}{d\tau}d\tau. \tag{6.28}$$

Integrating by parts:

$$\epsilon (t) = \sigma (t) J (0) + \int_0^t \sigma (\tau) \frac{dJ (t - \tau)}{d(t - \tau)} d\tau. \tag{6.29}$$

The integrals may be written as follows:

$$\epsilon (t) = \int_0^t J (t - \tau) \frac{d\sigma (\tau)}{d\tau} d\tau, \tag{6.30}$$

and

$$\epsilon (t) = \int_0^t \sigma (\tau) \frac{dJ (t - \tau)}{d (t - \tau)} d\tau, \tag{6.31}$$

where the initial term in equations (6.28) and (6.29) is now included in the integrals. Further, they may be expressed in terms of Stieltjes integrals; for instance equation (6.30) may be written as

$$\epsilon (t) = \int_0^t J (t - \tau) \, d\sigma (\tau), \tag{6.32}$$

a form that combines summations and Riemann integrals into one equation; see Appendix 2 in Jordaan (2005).

Integrals of the kind just described are of the form

$$h (t) = \int_0^t g (t - \tau) \frac{df (\tau)}{d\tau} d\tau = \int_0^t g (t - \tau) \, df (\tau), \tag{6.33}$$

and termed hereditary integrals, or convolution integrals. In the latter terminology, the integral is a convolution of $f (\cdot)$ and $g (\cdot)$. We can superimpose stresses resulting from strain increments rather than strains resulting from stress increments, to obtain

$$\sigma (t) = \int_0^t R (t - \tau) \frac{d\epsilon}{d\tau} d\tau, \tag{6.34}$$

or, in Stieltjes form

$$\sigma (t) = \int_0^t R (t - \tau) \, d\epsilon (\tau). \tag{6.35}$$

We can write the convolutions above using "composition products" in the following way (Fung, 1965):

$$h = g * df = f * dg,$$

$$J * d\sigma = \sigma * dJ,$$

and

$$R * d\epsilon = \epsilon * dR.$$

The Laplace transform of $h (t) = \mathcal{L} (h) = \overline{h}$, and in equation (6.33) this is the product of the transforms of f and g as follows:

$$\overline{h} (s) = \overline{f} (s) \cdot \overline{g} (s). \tag{6.36}$$

This is the convolution theorem. The proof appears in the standard texts noted earlier. As a consequence, for step inputs of σ and ϵ at $t = 0$

$$\overline{\epsilon} = s\overline{J}\overline{\sigma} \tag{6.37}$$

and

$$\overline{\sigma} = s\overline{R}\overline{\epsilon}. \tag{6.38}$$

Consequently

$$\overline{J}(s)\,\overline{R}(s) = \frac{1}{s^2}. \tag{6.39}$$

By taking the inverse of equation (6.39) and noting that $\mathcal{L}(t) = 1/s^2$, it is found that

$$\int_0^t J(t-\tau)R(\tau)\,d\tau = \int_0^t R(t-\tau)J(\tau)\,d\tau = t. \tag{6.40}$$

The mathematical background presented in this section is fundamental in viscoelastic theory and is used in Section 6.11 in introducing Schapery's Modified Superposition Principle, a theory of great utility in ice mechanics and generally in time-dependent fracture and damage mechanics.

6.7   Retardation and Relaxation Spectra

In equation (6.21b) we defined the retardation time, $t_R = \mu_k/E_k$, and used it as follows in a Kelvin unit under constant stress σ_0 applied at $t = 0$:

$$\epsilon(t) = \frac{\sigma_0}{E_k}\left[1 - \exp\left(-\frac{t}{t_R}\right)\right] \text{ for } t \geq 0. \tag{6.41}$$

Viscoelastic strain is most generally represented by a series of springs and dashpots in order to obtain the most detailed analytical description of observed stress–strain relationships for the material being studied. The most general canonical forms are shown in Figure 6.9; these are derived in Appendix A. We are studying linear theory at present but they can also be used in nonlinear theory with nonlinear springs and dashpots. In the case of a series of Kelvin units:

$$\epsilon(t) = \sigma_0 \sum_i \frac{1}{E_k^i}\left[1 - \exp\left(-\frac{t}{t_R^i}\right)\right], \tag{6.42}$$

where E_k^i is the ith elastic compliance, and $t_R^i = \mu_k^i/E_k^i$ is the ith retardation time. We can write this as

$$\epsilon(t) = \sigma_0 \sum_i \Delta\varphi\left(t_R^i\right)\left[1 - \exp\left(-\frac{t}{t_R^i}\right)\right], \tag{6.43}$$

where

$$\Delta\varphi\left(t_R^i\right) = \frac{1}{E_k^i} = \frac{1}{\mu_k^i}t_R^i. \tag{6.44}$$

(a)

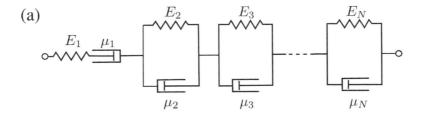

(b)

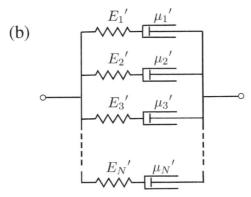

Figure 6.9 Canonical forms of viscoelastic models. (a) Kelvin chain. (b) Maxwell chain.

If the number of Kelvin units increases indefinitely, then we can write

$$\Delta\varphi\left(t_R^i\right) \rightarrow \frac{d\varphi}{dt_R}dt_R, \quad \text{and} \tag{6.45}$$

$$\epsilon(t) = \sigma_0 \int_0^\infty \frac{d\varphi(t_R)}{dt_R}\left[1 - \exp\left(-\frac{t}{t_R}\right)\right]dt_R \tag{6.46}$$

$$= \sigma_0 \int_0^\infty M(t_R)\left[1 - \exp\left(-\frac{t}{t_R}\right)\right]dt_R. \tag{6.47}$$

In this equation, $M(t_R)$ is the "retardation spectrum". In a full linear analysis, the complete spectrum of units contributing to the compliance might be considered. This may be a difficult thing to achieve; generally some approximation appropriate to the purpose at hand is sufficient.

Similar results can be found for the stress σ under imposed constant strain ϵ_0 for a single Maxwell unit:

$$\sigma(t) = \epsilon_0 E_m \exp\left(-\frac{t}{t_L}\right), \tag{6.48}$$

where E_m is the elastic modulus and $t_L = \mu_m/E_m$ is the relaxation time. For a series of Maxwell elements,

$$\sigma(t) = \epsilon_0 \sum_i E_m^i \exp\left(-\frac{t}{t_L^i}\right), \tag{6.49}$$

where the E_m^i are elastic moduli and the t_L^i are the relaxation times. We may write

$$\Delta\psi\left(t_l^i\right) = \Delta E_m^i \rightarrow \frac{d\psi\left(t_L\right)}{dt_L}dt_L,\tag{6.50}$$

and consequently

$$\sigma\left(t\right) = \epsilon_0\int_0^\infty \frac{d\psi\left(t_L\right)}{dt_L}\exp\left(-\frac{t}{t_L}\right)dt_L\tag{6.51}$$

$$= \epsilon_0\int_0^\infty N\left(t_L\right)\exp\left(-\frac{t}{t_L}\right)dt_L.\tag{6.52}$$

The quantity $N\left(t_L\right)$ is the "relaxation spectrum". (Figure 6.9 introduced the generalized chains in canonical form.) The equations may be written in terms of the differential operators $\mathbb{R}$ and $\mathbb{Q}$ (see Flügge, 1967, for example):

$$\mathbb{R}\sigma = \mathbb{Q}\epsilon.\tag{6.53}$$

The linear theory described can be extended to consideration of nonlinear springs and daspots as shown by Schapery (1966, 1968, 1969).

To the author's knowledge, there has been no systematic study aimed at determining the full spectrum of ice viscoelastic response. Xiao (1997) used three Kelvin units, including the effects of nonlinearity in the dashpots and damage, to model strain under multiaxial stress conditions. In the next section we investigate the use of nonlinear dashpots in Kelvin elements, and we use Andrade's equation (related to Sinha's), for which we deduce the spectrum for that particular relationship.

6.8   Use of Nonlinear Dashpots: Spectrum for Andrade's Relationship βt^b

In Section 2.4, we discussed Sinha's equation for creep of ice, in particular equations (2.5) and (2.6). For modelling of delayed elasticity, the use of a Kelvin model seems a natural approach. The spring in Sinha's model is linear, but the exponent is raised to the power $(a_Tt)^b$. This is not a linear term as occurs in the Kelvin model. The Sinha relationship gives a more rapid rise in delayed elastic strain than would be obtained with a linear term in the exponent. In this section, we discuss two ways of dealing with this. The first is to consider the viscosity in the dashpot of the Kelvin model to be stress-dependent; the second is to consider further the approximation embodied in the expression (2.7).

We have already seen that the dashpot corresponding to the "flow" term of Glen's law is nonlinear as shown in equation (6.13) with $n \simeq 3$. If the dashpot in the Kelvin element of the Burgers model is nonlinear $(m > 1)$, where m is the power to which the dashpot stress σ_d is raised, i.e. $\left(\sigma_d^m\right)$, this will result in a more rapid rise in the delayed elastic strain with time, as desired. We use this approach for the dashpot in the Kelvin element:

$$\dot\epsilon_k = \dot\epsilon_0\left(\frac{\sigma_d}{\sigma_1}\right)^m = \frac{\sigma_d}{\mu_k\left(\sigma_d\right)},\tag{6.54}$$

where, as noted, σ_d is the stress in the dashpot, ϵ_k the strain, and $\mu_k(\sigma_k)$ the stress-dependent viscosity in the Kelvin element. We may then deduce that the strain response of the Kelvin element to stress σ applied at $t = 0$, and held constant, to be

$$\epsilon_k(t) = \frac{\sigma}{E_k}\left[1 - \exp\left(-\int_0^t \frac{E_k}{\mu_k(\sigma_d)}d\tau\right)\right], \tag{6.55}$$

where E_k is the Kelvin element spring stiffness. The dashpot viscosity becomes effectively a function of time. This approach was used in Jordaan and McKenna (1988) with some success.

In a situation involving time-stepping where the external stresses applied to a material point are changing, the stresses in the dashpot must of course be carefully tracked, and can also be used to obtain the current viscosity. But this eliminates the need to track past history of stress and strain, making the process tractable. Xiao and Jordaan (1996) showed excellent agreement between the approach just described for ice tested under moderate confinement of 10 MPa. The relationship has been used in several studies for instance Xiao (1997), Singh (1993), Li (2002) and Turner (2018). The method is related to that of Schapery (1968) (see Appendix A) and the use of reduced time or "pseudotime":

$$\psi(t) = \int_0^t \frac{dt'}{a_D}; \psi(\tau) = \int_0^\tau \frac{dt'}{a_D}. \tag{6.56}$$

Replacing time t with pseudotime $\psi(t)$ achieves the same objective of addressing nonlinearity of the dashpots. The quantity a_D is in effect a viscosity. In general, different material points might have different stress histories and therefore different viscosities.

We now consider the second method based on the approximation (2.7), where we introduced the relationship βt^b for representing delayed elastic strain with $b \simeq 1/3$. This was shown to represent well Sinha's delayed elastic formulation for small t, as shown in equation (2.6). We shall now derive the retardation spectrum for this kind of relationship (see also Findley et al., 1976). We first equate the strain to that in equation (6.47):

$$\beta t^b = \int_0^\infty M(t_R)[1 - \exp\left(-\frac{t}{t_R}\right)dt_R], \tag{6.57}$$

from which we aim to deduce the distribution of retardation times $M(t_R)$ for Andrade's equation. First, we differentiate equation (6.57) with respect to t:

$$\beta b t^{b-1} = \int_0^\infty M(t_R)t_R^{-1}\exp\left(-\frac{t}{t_R}\right)dt_R. \tag{6.58}$$

Now we substitute $t_R = 1/u$, with $dt_R = -\left(1/u^2\right)du$, and suppose that $M(t_R)$ takes the power-law form ct_R^d. Then

$$\beta b t^{b-1} = ct^d \int_0^\infty (tu)\exp(-tu)d(tu) \tag{6.59}$$

$$= ct^d\Gamma(-d), \tag{6.60}$$

where $\Gamma(x) = \int_0^\infty w^{x-1} \exp(-w)\, dw$ is the gamma function. We then conclude that the power-law form is appropriate, and that $d = b - 1$, with $c = \beta b / \Gamma(1 - b)$. As a result

$$M(t_R) = \frac{\beta b t_R^{b-1}}{\Gamma(1 - b)}. \tag{6.61}$$

Generally we are interested in $b < 1$, and in particular $b \simeq 1/3$. The distribution of retardation times of equation (6.61) tends to infinity as $t_R \to 0$, and decreases with increasing t_R. In log-space, a straight line is obtained. A broad range of distribution times is covered.

The Andrade form just discussed has been used in a variety of studies regarding ice, e.g. Schapery (1991a), Jordaan et al. (1992), Singh (1993) and Kavanagh (2018).

6.9   Stress and Strain Redistribution: Correspondence Principle

Stress and strain can redistribute significantly in structures composed of viscoelastic material. This will be a key factor in the analyses to follow. Very significant local-ization of stress and strain arises in (for example) compressive ice failure, associated with viscoelastic behaviour coupled with the effects of microstructural change. It is helpful to understand the factors that result in this redistribution. But redistribution does not always happen. First, we consider some structures in which it does not occur in linear viscoelasticity, for example a simply supported (horizontal) beam on rigid supports with constant loading. Statics would dictate that under statically determinate conditions, the reactions remain the same, as do the stresses. All strains and deflec-tions will be proportional to the creep compliance: $\epsilon(x) = \sigma(x) J(t)$, where $\sigma(x)$ is the stress at point x in the body. This is the first form of a correspondence principle: replace the value E by $1/J(t)$ to obtain the strain or deflection of the beam at time t. The correspondence is between the elastic solution using E and the viscoelastic solu-tion found by replacing E by $1/J(t)$. The latter value is referred to as the "effective modulus". If the system under consideration is statically indeterminate, for example a beam fixed at both ends, nothing in the preceding discussion alters. The use of an "effective modulus" can also assist in solving the convolution integrals in viscoelastic theory, if appropriate approximations can be found. This is done in fracture analysis (Chapter 9).

The situation does indeed alter when the viscoelastic body interacts with another body with different stress–strain behaviour. An example is a viscoelastic beam on two simple supports at the ends, and with an elastic spring support at the centre. The beam and spring are initially unloaded; when loaded, the beam deflects and starts to load the spring. Then, as the beam creeps, it will further transfer load over time to the central spring, and load and stress redistribution take place. See also Flügge (1967). Kavanagh (2018) studied the redistribution in a beam under four-point loading with and without a spring support at the centre; the beam was also used in his fracture analysis and testing. Stress analysis for linear viscoelastic materials can be conducted using the following principle. An interested reader might wish to research this aspect further; the principle is well described in the standard works.

Elastic-Viscoelastic Correspondence Principle. Consider a linearly viscoelastic body with loads or prescribed displacements applied at time $t = 0$. The Laplace-transformed solution can be obtained from the elastic solution by replacing the elastic modulus by $1/[s\overline{J}(s)]$, where $\overline{J}(s)$ is the Laplace transform of $J(t)$. As noted, this leads to the Laplace transformation of the viscoelastic problem. The boundary constraints for this method must be independent of time or separable functions of space and time. The solution in time is found by inverting the Laplace-transformed solution.

This approach is used later in fracture analysis (Section 9.4). For this purpose, the extension of theory by Graham (1968) is needed, since there are moving boundaries associated with crack growth. This analysis was first based on linear viscoelasticity and then extended by Schapery to the nonlinear case using MSP (introduced in Section 6.11) together with a new correspondence principle. The solutions also rely on the use of an effective modulus.

Stress redistribution in nonlinear viscoelastic materials is extensive. Consider, for example, a nonlinear viscoelastic beam subjected to a constant bending moment. Assume a power-law creep relationship $\dot{\epsilon} \propto \sigma^n$, say with $n = 3$, and with an initial linear elastic response. On the basis of "plane sections remain plane" (i.e. ignoring shear deformation for the illustration), the initial stress distribution will be linear, following the strain distribution. The strain rates in the outer fibres will then be much higher than near the centre as a result of the higher stress there and the associated nonlinearity. To preserve equilibrium and the linear strain distribution, the stress drops near the outer fibres, and increases nearer the neutral axis. This continues, resulting in a distinctly nonlinear stress distribution, tending towards the rectangular shape associated with plastic bending. In the case of damaged ice, microstructural changes occur (Figure 2.1) in the formation of the layer in compressive failure. The damage has the effect of very significantly increasing the creep compliance. This results in substantial localization of deformation into the layer itself, and creates the rather exceptional failure mode of ice in compression.

6.10 Early Damage Mechanics and Experiments

The idea of a state variable representing damage has its origin in the seminal work of L. M. Kachanov (1958). This is one of the guiding concepts used in the present work. L. M. Kachanov (1958) and Rabotnov (1969) represented the damage state of materials containing distributed cavities in terms of internal state variables, and then established equations to describe their evolution and the resulting mechanical behaviour. The simplest way to characterize a damaging material is by means of a single internal variable, for instance taking the original cross-sectional area A_0 as having been reduced by a smaller area A. In this section and the next, we deal with compressive states so that we take these as being positive, with associated contractive deformations as being positive also. Under uniaxial loading, say P, the initial stress in the body (without damage) is $\sigma = P/A_0$. The damage variable λ is defined as

$$\lambda = \frac{A}{A_0}, 0 \le \lambda \le 1. \tag{6.62}$$

The "effective stress" σ_e is defined as

$$\sigma_e = \frac{P}{A_0 - A} = \frac{\sigma}{1 - \lambda}. \tag{6.63}$$

The strain ϵ in the material is given by

$$\epsilon = \frac{\sigma_e}{E_0} = \frac{\sigma}{E_0(1 - \lambda)} = \frac{\sigma}{E}, \tag{6.64}$$

with E representing the "damaged" or "effective" elastic modulus, and where E_0 is the elastic modulus of the virgin undamaged material. We may write

$$E = E_0(1 - \lambda). \tag{6.65}$$

The "effective elastic modulus" is given by this value, reduced from the original value, leading to modified stress and strain in the material.

A parameter varying from 0 to 1 indicating the degree of damage, not necessarily the reduced area, has been used successfully in studies of metal failure (Kachanov, 1958; Leckie, 1978). With regard to applications in ice mechanics, the writer was attracted to this approach in the 1980s. Early work was conducted by Sjölind (1987) and Karr and Choi (1989), who developed three-dimensional viscoelastic models with microcracking as the damage variable. The "damage" in the early studies concentrated on the effect of microcracking as "damage" to the ice, with consequent effects upon the elastic and creep response. In our studies, we initially considered uniaxially stressed specimens (Jordaan and Mckenna, 1988, 1991; Stone et al., 1989; McKenna et al., 1990; Jordaan et al. 1992; Xiao and Jordaan, 1996).

Microcracks in the material can lead to locally increased stresses and strains; for instance Budiansky and O'Connell (1976) studied the strain energy loss resulting from the nucleation of individual cracks. For an isotropic array of similarly shaped flat circular cracks, the isotropic damage parameter is simply related to the crack density by $\lambda = a^3 N$, where a is the radius of crack surface and N is the crack density. Their results include interaction between cracks but do not account for traction across crack surfaces; all cracks remain open. M. Kachanov (1993) studied this question and found that the effect on compressive elastic moduli to be reduced, as compared to the tensile case. A 3D solution was proposed for non-interacting cracks with an isotropic random distribution. These are stated in Xiao and Jordaan (1996) and we quote here the effect on Young's elastic modulus for illustration:

$$\frac{E}{E_0} = (1 + C_1 S)^{-1}, \tag{6.66}$$

where

$$C_1 = \frac{16(1 - v_0^2)(1 - 3v_0/10)}{9(1 - v_0/2)}, \tag{6.67}$$

with ν denoting Poisson's ratio. In this case the damage $S = N$, the number of cracks per unit volume. The results are valid for low and high crack densities.

The effect of cracks on the steady-state creep rate for the two-dimensional case was examined by Weertman (1969) using dislocation theory. Approximate solutions were given for materials obeying the power law creep equation. His result for low crack density was used by Sinha (1988, 1999); see also Xiao and Jordaan (1996). For high crack densities, Weertman noted that creep rates

$$\dot{\epsilon} = A\sigma^n \tag{6.68}$$

for the material without cracks would be of the order of

$$\dot{\epsilon} \sim A\sigma^n \left(a^2 N_1 \right)^{n+1}, \tag{6.69}$$

for the cracked material, where a is the half-crack-size, and N_1 is the crack density (number of cracks per unit area). In other words, there is an enhancement of the order of $\left(a^2 N_1 \right)^{n+1}$. This factor can be quite considerable. The power-law remains in effect. In ice failure, there inevitably will be high crack densities, and we consider in the following that N is the crack density per unit volume. Thinking also of higher powers of N, it was proposed to use an exponential term as follows:

$$\dot{\epsilon}_N = \dot{\epsilon} \exp(\alpha N), \tag{6.70}$$

where $\dot{\epsilon}$ is the reference rate of equation (6.68) without cracks, and α is a constant, and where

$$N = \int_0^t \dot{N} d\tau. \tag{6.71}$$

Next, we consider the rate of microcrack formation, $\dot{N}$. Cracks were observed to grow rapidly, and then stop, with the orientation being approximately parallel to the principal (uniaxial) compressive stress. Initially, we considered the use of rate theory (Section 2.8) expressed in exponential form, but then simplified this (McKenna et al., 1990) to

$$\dot{N} = \dot{N}_0 \left(\frac{\sigma - \sigma_c}{\sigma_0} \right)^q \tag{6.72}$$

where σ is the uniaxial stress, $\dot{N}_0$ is a rate constant, σ_c is the threshold value of stress (quite small), $\sigma_0 = 1$ MPa is a normalizing stress and q is a constant.

The analysis above is well supported by experimental work. Initially, only uniaxial tests were possible; these are reported in Jordaan et al. (1992), and further uniaxial tests as well as tests under moderate confining tests are reported in Stone et al. (1989). (Appendix B contains details of specimen preparation and procedures). The tests described in Jordaan et al. (1992) were designed to investigate the deformation of ice under load, and the influence of cracks and level of damage on the consequent creep response, and to obtain the relevant material constants for constitutive modelling. To study the effect of microcracking, some specimens were kept intact (undamaged specimens), while others were loaded at a constant strain rate of 1×10^{-4} s^{-1} to a maximum

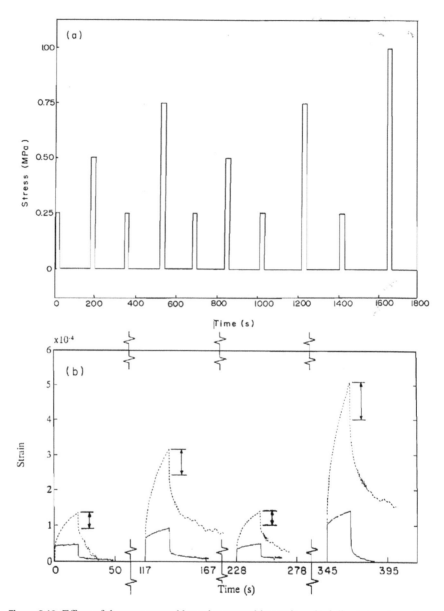

Figure 6.10 Effect of damage caused by prior stress history in uniaxially stressed specimens (Jordaan et al., 1992). (a) Stress Histories. (b) Solid lines show strains in undamaged specimens, dotted lines show strains in damaged specimens. With permission from Canadian Science Publishing.

strain of 0.02. These were termed "damaged" specimens. Specimens with both treatments (intact and damaged) were then subjected to 20 s constant loads at pressures of 0.25, 0.50, 0.75 and 1.00 MPa. Figure 6.10 shows typical results. It can be seen that the prior stress history has a very strong effect on the material response.

A viscoelastic damage model was also developed to estimate the consequences of microcracking on the stress–strain response, along the lines of the theory above, using equation (6.66), simplified to

$$E = E_0 \left(1 - \frac{16}{9}\lambda_N\right);$$ (6.73)

(see Budiansky and O'Connell, 1976; McKenna et al., 1990, and discussion in Jordaan et al., 1992), where

$$\lambda_N = a^3 N,$$ (6.74)

in which a is the average radius of flat circular cracks, and N is, as before, the crack density. When the crack size is of the order of grain size, as is the case for ice,

$$\lambda_N = \frac{N}{8N_g},$$ (6.75)

where N_g is the volumetric density of one crack per grain; thus

$$N_g = d_g^{-3},$$ (6.76)

with d_g representing the average grain diameter. Both flow (Maxwell element) and delayed elastic (Kelvin) terms were included in the analysis, using nonlinear dashpots in each. The flow rate in the dashpots was enhanced by a factor $\exp\left(\alpha N/N_g\right)$, $\alpha =$ constant, as a result of damage. Reasonable agreement of theory with experiment was found. (See Figure 8 in Jordaan et al., 1992).

The tests of Stone et al. (1989) were conducted under uniaxial conditions and under small confinements (2.5 and 5.0 MPa) in a triaxial cell, at constant strain rates of $10^{-3}, 10^{-4}$ and 5×10^{-5} s^{-1}. Specimens were strained until after peak stress had been reached, and then unloaded to investigate the effect of the damage that had occurred in the initial loading. Figure 6.11 shows typical results. The damaged material eventually reaches a plateau at the given rate, much like a plastic material, but the plateaux are reached after a period of damage occurrence which eventually stabilises; the plateaux will be different for different rates, so that there is no unique plateau. Damage was assessed using equation (6.65), based on measurements of elastic modulus, thereby giving measured values of λ. These were obtained by calculating the elastic modulus on re-loading after the initial loading cycle, and comparing with the value on initial loading. This was repeated up to 4 times. These tests showed λ values giving a reduction of the order of 50–60% in elastic modulus. At the same time, the creep magnification $\exp(\beta N)$ of equation (6.70) could be as high as 400. Creep enhancement was seen to be likely to be the dominant effect with regard to the effect of damage. The test results supported the notion that the power n in equation (6.68) is approximately constant, in the range 3–4, with and without damage; see also Duval et al. (1991) and the work of Meyssonnier and Duval (1989).

Using equation (6.72) with $\sigma_c = 0$, and equation (6.71), we can derive the damage parameter D_N related to microcrack formation as

$$D_N = \frac{\dot{N}_0}{N_g} \int_0^t \left(\frac{\sigma}{\sigma_0}\right)^q d\tau,$$ (6.77)

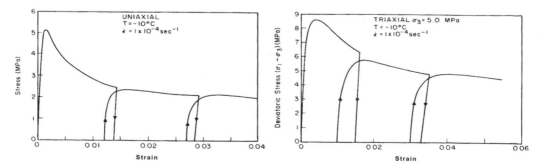

Figure 6.11 Repeat loadings of uniaxially and triaxially loaded specimens showing effect of prior stress history, after Stone et al. (1989). With permission from POAC.

with associated enhanced creep rate

$$\dot{\epsilon} = \dot{\epsilon}_0 \left(\frac{\sigma}{\sigma_0} \right)^n \exp\left(\alpha D_N\right). \tag{6.78}$$

In further work, the damage state included other microstructural changes observed in testing, in particular recrystallization, and is written as S, rather than D_N; based on the preceding:

$$S = \int_0^t f(p, \ldots) \left(\frac{\sigma}{\sigma_0} \right)^q d\tau, \tag{6.79}$$

where $f(\ldots)$ is a function of pressure p, and possibly other variables, and q is a general constant reflecting a variety of processes, including recrystallization. This was found to be consistent with Schapery's measure of damage.

6.11   Schapery's Use of Modified Superposition Principle (MSP) in Ice Mechanics

In Schapery (1991a), the "modified superposition principle" (MSP) in viscoelastic analysis of ice was introduced. This is an elegant piece of theory, which has great potential for further applications in ice mechanics, as it is possible to deal with non-linearity and damage in a tractable manner. It has been used in many areas including for example biomechanics (Lakes and Vanderby, 1999), and is here summarized with the application to ice in mind. We deal here with the uniaxial case (a full treatment is given in Chapter 9 and Appendix A). We start with linear material response; the equation is of the familiar single-integral form with the strain ϵ under uniaxial stress σ given by

$$\epsilon = \int_0^t D(t - \tau) \frac{d\sigma}{d\tau} d\tau, \tag{6.80}$$

where $D(\cdot)$ is a linear creep compliance. The inverse of this equation is

$$\sigma = \int_0^t E(t - \tau) \frac{d\epsilon}{d\tau} d\tau. \tag{6.81}$$

Using Laplace transforms as in equation (6.39):

$$\overline{D}(s)\,\overline{E}(s) = \frac{1}{s^2}. \tag{6.82}$$

For a power-law creep function in time, as in for example Section 6.8 and equation (2.7), of the form

$$D = D_1 t^n, \tag{6.83}$$

the Laplace transform is

$$\overline{D}(s) = D_1 \frac{\Gamma(n+1)}{s^{n+1}}, \tag{6.84}$$

where Γ is the gamma function. See also Lakes (2009). Using equation (6.39), we have

$$\overline{E}(s) = \frac{s^{n-1}}{D_1 \Gamma(n+1)}. \tag{6.85}$$

We can invert this and obtain

$$E(t) = \frac{t^{-n}}{D_1 \Gamma(n+1)\,\Gamma(1-n)}. \tag{6.86}$$

Since

$$\Gamma(n+1)\,\Gamma(1-n) = \frac{n\pi}{\sin(n\pi)}, \tag{6.87}$$

we have also that

$$E(t) = \frac{\sin(n\pi)}{D_1 n\pi} t^{-n}, \tag{6.88}$$

and that

$$ED = \frac{\sin(n\pi)}{n\pi}. \tag{6.89}$$

The theory so far is within the domain of linear viscoelasticity.

The very useful aspect explored by Schapery in many ingenious ways is that the basic formulation may be extended to nonlinear materials, and also to the consideration of fracture and damage as well as to multiaxial stress states. We first progress to the consideration of nonlinear relationships, and a departure from conventional analysis is propounded. Nonlinear materials are analysed using

$$\epsilon(t) = \int_0^t D(t-\tau)\,\frac{\partial\psi}{\partial\tau}\,d\tau, \tag{6.90}$$

where $\psi(\sigma)$ is a nonlinear function of stress, and

$$\frac{d\psi}{d\tau} = \frac{d\psi}{d\sigma}\frac{d\sigma}{d\tau}. \tag{6.91}$$

The function $D(\cdot)$ is still a linear compliance. For example, if we wish to analyse a material following power-law viscous creep,

$$\frac{\dot{\epsilon}}{\dot{\epsilon}_0} = \left(\frac{\sigma}{\sigma_0}\right)^n, \tag{6.92}$$

we can substitute

$$D(t - \tau) = \frac{1}{t_v}(t - \tau),$$ (6.93)

where t_v is a constant, and consider a nonlinear elastic material

$$\psi(\sigma) = \left(\frac{\sigma}{\sigma_0}\right)^n.$$ (6.94)

Then, solving, we find that $t_v = 1/\epsilon_0$, and for stress σ_0 applied at $t = 0$

$$\epsilon(t) = \left(\frac{\sigma}{\sigma_0}\right)^n \frac{t}{t_v}.$$ (6.95)

For more general functions ψ, and using equation (6.90), for a creep test with stress input σ_0 at $t = 0$, and removed at $t = t_1$, then

$$\sigma(t) = \sigma_0[H(t) - H(t - t_1)],$$ (6.96)

where $H(\cdot)$ is the unit step function and then

$$\epsilon(t) = \psi(\sigma_0)[D(t) - D(t - t_1)]$$ (6.97)

for $t > t_1$. Findley et al. (1976, p. 229) describe use of the superposition principle and the MSP in some detail. The big departure from the usual theory of viscoelasticity is that the stress–nonlinearity is not included in the springs and dashpots, but in the second term, in ψ, the derivative of which we take in equation (6.90). Since the function $D(t)$ is not dependent on state variables and the value E_R is a constant, the theory imposes a condition that Poisson's ratio is constant during the process, certainly a reasonable approximation for the creep processes in ice.

It is important to bear in mind that the J integral, of fundamental importance in fracture theory, relies upon a potential function which can be a nonlinear function of stress. This is an important contribution by Schapery, because it allows a systematic study of time-dependent fracture using correspondence principles that are consistent with the new formulation that we are discussing. This is treated further in Chapter 9, where tensile states are considered in fracture analysis. The present analysis extends beyond nonlinearity to include damage:

$$\epsilon(t) = E_R \int_0^t D(t - \tau) \frac{d\varrho(\tau)}{d\tau} d\tau,$$ (6.98)

where $\varrho(\tau)$ reflects both nonlinearity with stress and damage:

$$\varrho = \varrho(\sigma, S_k).$$ (6.99)

The quantities S_k $(k = 1, 2, \ldots)$ reflect the effect of microcracking (or other microstructural change) on the strain ϵ. For proportional stressing, of which uniaxial stressing is a particular case, and given power law nonlinearity with stress as well as for crack growth rate, then we find that

$$\varrho = \left(\frac{\sigma}{\sigma_2}\right)^r g(S),$$ (6.100)

and

$$S = \int_0^t \left(\frac{\sigma}{\sigma_1}\right)^q f_1 d\tau, \tag{6.101}$$

where σ_1, σ_2, r and q are positive constants. See Appendix 1 for further details.

The similarity of equation (6.101) to equation (6.79) is noted. If we integrate the rate equation (6.72), with $\sigma_c = 0$, as in equations (6.77) and (6.79), our previously developed analysis of damage, carried out using intuitive physical reasoning, is closely related to equation (6.101). This is exploited in the further work described in Chapters 7 and 8. Schapery discusses the function $g(S)$ for enhancement of the creep rate:

$$g(S) = \exp(\beta_S S), \tag{6.102}$$

where β_S is a constant. Again, we observe the close relation of equation (6.70) to (6.102). The term E_R in equation (6.98) was introduced by Schapery as a free constant which he termed the "reference modulus", with units of stress. Our first encounter with Schapery's correspondence principles is found if we put $D = 1/E_R$, and noting that this is now an elastic material (possibly nonlinear), we have

$$\sigma = \sigma^R; \epsilon^R = \varrho(\sigma, S_k). \tag{6.103}$$

The superscript R denotes the reference elastic solution, which is used to derive its viscoelastic counterpart. The quantity $\epsilon^R = \varrho$ now represents the elastic strain in material with the same damage values as in the viscoelastic material. In Schapery's correspondence principle (Chapter 9 and Appendix A), $\sigma(t) \equiv \sigma^R$, and $\epsilon(t)$ is given by equation (6.98).

Schapery (1991a) derived constant stress and strain-rate equations for uniaxial stressing, including damage; in fact these had been derived using the Schapery theory by Harper (1986) with promising results. Schapery also analysed the results of Ashby and Duval (1985) in which the creep of ice is expressed in terms of the minimum creep rate. His results showed excellent agreement with other results and showed that damage is needed to explain the minimum creep rate and the subsequent ascent into tertiary creep. His analysis was based on an expression without the steady flow term. The results in Jordaan et al. (1992) suggest that the inclusion of this flow term would be a good and important addition in studying ice behaviour. Comparisons with tests involving loading and unloading, and with re-loading, are important in this regard.

7 Complex States of Stress and Triaxial Tests

7.1   Multiaxial States of Stress

It is clear that microstructural change, mainly microcracking in the early tests with low confinement, leads to significant changes in material response. It is also known that compressive failure in *hpzs* involved high pressures—up to 70 MPa contact pressures having been measured in the *hpzs*—together with widespread recrystallization. These had not been addressed in the earlier research. Part of the strategy was to formulate finite-element analyses so as to use the constitutive modelling—including the effects of microstructural change—in the finite element analysis. One of the main aspects needing investigation is the intense localization of ice pressure during ice crushing, a process that is not consistent with conventional mechanics—see Figure 2.1. Confining pressure will suppress microcracking, but the role of recrystallization was not clear. Relevant data on ice behaviour under high confining pressure were needed for further development of theory of ice failure in compression. Various positions within the zone are illustrated in Figure 7.1, which correspond to possible different confining pressures. It is clear that significant confining pressures must exist in these zones. As a result, several series of tests were conducted under triaxial stress conditions. These proved to be determining factors in understanding how failure might take place.

Our states of stress are largely compressive, so we take compressive stress and contractive strain as being positive. We assume in our analysis that our material can be treated as being isotropic (statistically): the specimens for testing were of carefully prepared granular ice. We use the division into hydrostatic pressure and volumetric strain, on the one hand, and deviatoric stress and strain, on the other. For stress and strain states σ_{ij} and ϵ_{ij}, we have the volumetric components p and e_v as

$$p = \frac{1}{3}\sigma_{ii}; e_v = \frac{1}{3}\epsilon_{ii}, \tag{7.1}$$

(repeated suffices denoting summation), with deviators expressed as

$$s_{ij} = \sigma_{ij} - p\delta_{ij}; e_{ij} = \epsilon_{ij} - e_v\delta_{ij}, \tag{7.2}$$

where δ_{ij} is the Kronecker delta = 1 if $i = j$, and = 0 otherwise. Generally, hydrostatic pressures result in change in volume, while the deviators relate to change in shape (shearing effects, as would be expected). Note that the first invariant I_1 of the stress

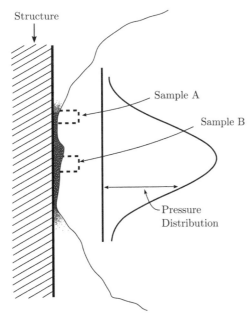

Figure 7.1 Schematic illustration of high-pressure zone at possible positions in the ice mass (A and B) and associated stress states for testing and simulation. With permission from Canadian Science Publishing.

tensor is equal to $(\sigma_{11} + \sigma_{22} + \sigma_{33}) = 3\sigma_v$, with a similar relation for strain. The von Mises (equivalent) stress s and strain e are defined as

$$s = \left(\frac{3}{2}s_{ij}s_{ij}\right)^{1/2} \; ; e = \left(\frac{2}{3}e_{ij}e_{ij}\right)^{1/2} . \tag{7.3}$$

These relationships are most useful. In considering the viscosity at a point in the continuum, it is natural to use them to formulate the mechanics. The principal stresses are written σ_1, σ_2 and σ_3, while the principal deviatoric stresses are written s_1, s_2 and s_3. Similarly, principal strains and principal deviatoric strains are written $\epsilon_1, \epsilon_2, \epsilon_3$ and e_1, e_2, e_3, respectively. Then we can write equation (7.3) as

$$s^2 = \frac{3}{2}\left(s_1^2 + s_2^2 + s_3^2\right) ; e^2 = \frac{2}{3}\left(e_1^2 + e_2^2 + e_3^2\right) . \tag{7.4}$$

Consider now the state of stress $\{\sigma_1, \sigma_2 = \sigma_3\}$, common in triaxial testing with $\sigma_1 > \sigma_3$, then we can use the relationship

$$s^2 = \frac{(\sigma_1 - \sigma_2)^2 + (\sigma_2 - \sigma_3)^2 + (\sigma_3 - \sigma_1)^2}{2} \tag{7.5}$$

to deduce that under the triaxial test conditions just noted,

$$s = \sigma_1 - \sigma_3. \tag{7.6}$$

Hence the stress deviator, or stress difference, applied to a specimen under pressure σ_3 is also the von Mises stress s. A special case is uniaxial stressing, $\sigma_1 > 0, \sigma_2 = \sigma_3 = 0$, in which case

$$s = \sigma_1. \qquad (7.7)$$

The first invariant of the deviatoric stress tensor, $J_1 = 0$, whereas the second invariant of the deviatoric stress deviator, written J_2, is given by

$$J_2 = \frac{1}{2}\left(s_1^2 + s_2^2 + s_3^2\right) = \frac{s^2}{3}. \qquad (7.8)$$

This forms the basis of the von Mises yield surface, governed by the criterion

$$J_2 = k^2 = \frac{s^2}{3}, \qquad (7.9)$$

where k is a constant. Since the von Mises yield criterion is independent of the first stress invariant, it is appropriate for the analysis of plastic deformation for ductile materials such as metals, where it has been found that failure is pressure-independent. The von Mises stress is used to predict yielding of materials under complex loading, possibly from the results of uniaxial tensile tests. In the latter case, $s = \sigma_1 = \sigma_y$, and therefore $k = s/\sqrt{3} = \sigma_y/\sqrt{3}$. The von Mises criterion can be represented in principal-stress space as a cylinder of radius $\sqrt{2/3}\sigma_y$. Inside the cylinder the material is elastic; outside, plastic flow occurs, following the normality principle. Pressure-dependence can be handled by using a surface with varying diameter, dependent on the hydrostatic pressure. This strategy only goes a small way towards dealing with the complexities of ice failure. The localization into a layer, and the associated change to the deformational response, do not emerge.

Returning to viscoelastic analysis, we continue with a Burgers body having stress-dependent dashpots in both the Maxwell and the Kelvin units (Xiao and Jordaan, 1996). The creep rates in the dashpots (based on the uniaxial case with stress σ) are as follows:

$$\dot{\epsilon}^c = \dot{\epsilon}_0^c \left(\frac{\sigma}{\sigma_0}\right)^n \qquad (7.10)$$

for the Maxwell dashpot, and

$$\dot{\epsilon}^k = \dot{\epsilon}_0^k \left(\frac{\sigma^k}{\sigma_0}\right)^m \qquad (7.11)$$

for the Kelvin dashpot, with σ^k denoting the stress in the dashpot. Typically, $\dot{\epsilon}_0^c = 1.76 \times 10^{-7}\ s^{-1}$ ($T = 263\ K$) is the viscous strain rate for unit stress, and where $n = 3$, $\sigma = 1$ MPa. See Xiao and Jordaan (1996) for further details. We now consider the multiaxial situation. Viscosities at a point are calculated using the von Mises stress s. The stress s^k in the Kelvin dashpot is calculated using

$$s^k = s - G_k e^k \qquad (7.12)$$

where G_k is the elastic stiffness of the Kelvin spring, and e^k is the accumulated delayed elastic strain:

$$e^k = \int_0^t \dot{e}^k\, d\tau. \qquad (7.13)$$

The relationship between the deviatoric strain e_{ij} and equivalent strain e is given by

$$e_{ij} = \frac{3}{2} e \frac{s_{ij}}{s}. \tag{7.14}$$

One can verify this by considering uniaxial stress σ_1, noting that in this case $s = \sigma_1$ and that $s_{11} = 2\sigma_1/3$. Stress-dependent strain rate is used in both Maxwell and Kelvin dashpot elements:

$$\dot{e}_{ij}^c = \frac{3}{2} \dot{e}^c \frac{s_{ij}}{s} = \frac{3}{2} \dot{\epsilon}_0 \left(\frac{s}{\sigma_0} \right)^n \frac{s_{ij}}{s}, \tag{7.15}$$

and

$$\dot{e}_{ij}^k = \frac{3}{2} \dot{e}^k \frac{s_{ij}}{s} = \frac{3}{2} \dot{\epsilon}_0^k \left(\frac{s^k}{\sigma_0} \right)^m \frac{s_{ij}}{s} = \frac{3}{2} \dot{\epsilon}_0^k \left(\frac{s - G_k e^k}{\sigma_0} \right)^m \frac{s_{ij}}{s}. \tag{7.16}$$

The dashpot viscosities can be written as

$$\mu_m = \left(\frac{s}{\dot{\epsilon}_0^c} \right) \left(\frac{\sigma_0}{s} \right)^n \tag{7.17}$$

and

$$\mu_k = \left(\frac{s^k}{\dot{\epsilon}_0^k} \right) \left(\frac{\sigma_0}{s^k} \right)^m \tag{7.18}$$

for the Maxwell and Kelvin units, respectively. The stress-dependent viscosities are $\propto 1/s^{n-1}$ and $1/\left(s^k \right)^{m-1}$, respectively; see also equations (6.13) and (6.54).

7.2 Review of Past Triaxial Testing

In mechanical testing, it is most important that the test device imposes the desired state of stress on the specimen. This has been extensively studied, for instance in testing concrete in uniaxial compression. In this case, the failure mode consists (as is the case for ice) of vertical cracks aligned in the direction of loading. During testing, the specimen expands laterally, following the Poisson effect. The loading platens will generally be of much stiffer material (steel) than the specimen, and they restrain the specimen at the ends. Practice in concrete testing moved away from short specimens (e.g. cubes) to longer cylindrical specimens, the reasoning being that the lateral restraint exercised by the loading platens would be minimized over the central portion of the specimen, which is then stressed closely to the desired state. The same reasoning has been applied to triaxial testing so that the standard triaxial cell, in which reasonably long cylinders rather than cubes are used, results in a uniform state of stress to an acceptable approximation over the central part of the specimen. Confining pressure is generally applied by a fluid (silicone oil, in our tests). This is the most common test apparatus used to characterize ice failure under combined stresses. The tests typically have stress states with $\sigma_1 > \sigma_2 = \sigma_3$.

With regard to the so-called "true triaxial" testing, in which the attempt is made to apply three independent stresses, the problem of obtaining a uniform state of stress is considerably increased. Indeed, the additional gain to be obtained from such testing is not major, and the use of "brush platens" serves to complicate the situation. These have been used, and in fact were initially designed, for concrete testing. There has been significant discussion of the existence of stress concentrations caused by these brush platens. Ice is very prone to localize into shear faults (see Section 7.6), and this results from local stress concentrations, sometimes in the ice itself, or caused by platen–specimen effects. The following statement appears in Schulson and Duval (2009): "Except for loading along paths of very low triaxial confinement,..., loading along all other triaxial paths, whether of low or high confinement, is terminated by shear faulting". Presumably this refers to the attempt to obtain a "failure envelope" with increasing stress; shear faulting did not occur in the majority of our constant strain-rate or constant stress testing. Weiss and Schulson (1995) suggested that failure in their tests was related to boundary conditions (damage localized at the ice–platen interface). Apart from the question of increasing stress to failure, two causes need investigation with regard to shear faulting (always found in Schulson's tests): localized stress concentrations associated with platen effects and, second, heterogeneities in the ice crystal structure. No results showing uniform damage development were obtained in their test series. It is noteworthy that shear faulting does not occur in ice–structure interaction, rather localization into a layer adjacent to the structure–ice interface. The present author has never observed the faulting suggested by Schulson in any indentation tests (large or small) or in any ice–structure interactions. Since ice is a viscoelastic material, the focus should be on a flow relationship, taking into account microstructural changes to the material. It is not possible to analyse ice failure with a fixed failure envvelope.

Distributed damage, often uniform spatially, was a feature that was consistently observed in the present investigation (to be outlined in Section 7.4). The tests were conducted largely under constant strain rate or constant stress conditions. In our tests, very carefully conducted on polycrystalline ice, uniform failure was observed in a large majority of tests (see for example Meglis et al., 1999). The tests were not aimed at obtaining a "failure envelope" but rather at formulating viscoelastic–damage constitutive relationships for ice. Localization will be discussed further in Section 7.6, but the "true triaxial" test results will not be reviewed further in the present work. This aside, past work on triaxial testing of ice has been reviewed by Melanson et al. (1999) and by Barrette (2001). Much past work has been aimed at developing a "yield surface" or "failure envelope" based on time-invariant plasticity theory. We shall not review these studies in any depth but note that it is considered futile to try to represent a material such as ice in this restricted manner. We can obtain an infinity of peak (yield) stresses merely by altering the strain rate or rate of loading or the state of damage (e.g. Figure 6.11). Many past studies were restricted to relatively low confining pressures or shear stresses. Values of hydrostatic pressure up to about 60–70 MPa are needed for the present purposes. It was also considered most important to monitor the microstructure and changes thereto during the deformation process.

Simonson et al. (1975) tested laboratory grown, polycrystalline ice at −10°C up to 200 MPa hydrostatic pressure at constant strain rate. They found an increase in strength with increasing strain rate, and a decrease in strength with increase in hydrostatic pressure (but note the high range of hydrostatic pressure). An increase of pressure alone (no deviator) resulted in melting of ice at 100 MPa. Jones (1978, 1982) tested laboratory-grown, granular, freshwater ice up to 85 MPa confinement, with the strain rate controlled. Strength was found to increase up to 25–30 MPa confinement but to decrease slightly with further increase in confinement. This trend was found to be more evident for higher strain rates. The stress exponent n for creep was found to be higher for unconfined than for confined ice. Confinement induced ductile (as opposed to brittle) deformation. Jones (1982) showed that the strength of ice is pressure-dependent at strain rates above 10^{-5} s^{-1}, in the range where crack nucleation is suppressed. Jones (1982) also discusses the effects of pressure melting as a result of high stresses at the grain boundaries, causing softening of the material at higher pressures.

Nadreau and Michel (1986a,b) tested laboratory-grown granular freshwater and saline ice, as well as iceberg ice. Confining pressures up to 70 MPa were used. It was found that Glen's exponent n decreased with an increase in confinement. Increase of maximum shear stress up to 15–20 MPa confinement was found, followed by decrease at higher confinement. Microcracking and its effects on the stress–strain behaviour of ice at varying confining pressures were reported by Rist et al. (1994). Moderate confinement, while reducing the extent of microcracking, was reported to have little effect on the shear strength of the ice. Others have investigated the effects of constant strain rate or constant deformation rate at high rates (Rist et al., 1988; Kalifa et al., 1992). Kalifa et al. (1992) reported the pressure dependence of microcracking as well as a change in failure of the ice from brittle to ductile with the application of confinement. Only a few studies have focused on characterizing the microstructural changes to the samples during testing. Jacka and Maccagnan (1984) found that uniaxially loaded test specimens showed a systematic change in both grain-size and grain orientation with increasing strain.

The main features of the results noted above were pressure-hardening at lower confinements, followed by softening at higher confinements. A programme was devised to assist in the analysis of the mechanics of ice compressive failure, with high confinements and shear values (Section 7.4). First, a brief review of the initial efforts in mechanics involving viscoelasticity and damage using finite element methods is given.

7.3 Initial Finite Element and Damage Analyses

The first finite element analyses of the present programme are summarized in McKenna et al. (1990) and in Xiao (1991). These focussed on indentation of ice surfaces, with a flat indentor against an ice wedge (McKenna) and with a spherical indentor (Xiao). The latter was based on the slow medium scale spherical indenter tests (Frederking et al., 1990). These analyses used the damage model described in Section 6.10. McKenna's work also included consideration of friction across closed cracks, based on the work of Horii and Nemat-Nasser (1983), whose results were also considered by Xiao. Ice creep is modelled as deviatoric, and a small dilatation term

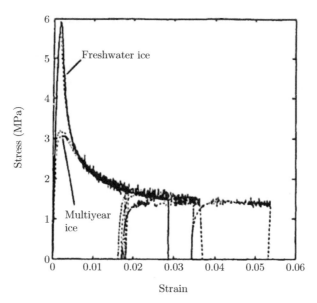

Figure 7.2 Laboratory and field ice compared; tests at strain rate 10^{-4} s^{-1}. Two specimens each of laboratory and multiyear field ice were tested at constant strain rate. A series of unloading and reloading cycles showed essentially the same response except for the initial peaks, which were higher for the laboratory specimens. From Xiao (1991), reproduced with permission.

was added. The viscosity coefficient is a function of the von Mises stress, as described in Section 7.1. The model is essentially a Burgers model, with stress- and damage-dependent dashpots, with linear springs, also damage-dependent. Both McKenna and Xiao included consideration of dilatation rate $\dot{e}_v$ resulting from crack formation. This effect is small, and it was recognized that creep strains dominated the results, especially as a result of the damage terms. It was also realized in these analyses that the effect of higher pressures on damage had not been examined and the results were appropriate for cases with smaller pressures.

Xiao (1991) analysed mutliyear ice samples that had been transported from the arctic in the region of the Hobson's Choice indentation tests. He conducted several uniaxial tests on the multiyear ice (labelled "intact" in his thesis) and also subjected this ice to damage cycles (Section 6.10). Figure 7.2 shows a comparison of the stress response to a strain rate of 10^{-4}s^{-1} of multiyear field ice and freshwater laboratory ice, of the kind commonly used in the present programme of work. There is a clear difference, and Xiao conducted a detailed analysis of parameters that would be suitable for analysing multiyear field ice from the area of the indentation tests. A comparison was made of the strain components for freshwater and multiyear ice. This included revision of creep rates and elastic parameters for the Burgers model. He also carried out tests on the effects of damage on the strains in the multiyear field ice and modelled the field situation by calculating an "initial damage" that could be applied to results from laboratory-grown ice to simulate field ice.

Xiao first made a comparison of the axisymmetric finite element model of indentation in ABAQUS using the User Material Subroutine (UMAT) with a theoretical elastic solution described by Sneddon (1964), and found very good agreement (Xiao, 1991). He then analysed two field indentation tests from the first Hobson's Choice series, NRC1 and NRC2, with indentor speeds of 0.3 and 2.5 mms^{-1}, respectively. The stresses from these slower tests were relatively small compared to tests involving crushing. The agreement between test and theory was found to be excellent (Xiao et al. 1990). The spread of the damage zone found in the analysis was found to be typical of the slower tests, as will be demonstrated further in Chapter 8. Xiao also studied pressure hardening; reference is made to Xiao and Jordaan (1996). At that stage in the investigation, there had been no analytical results obtained that supported the rapid declines in load observed in the faster medium-scale tests, but progress had been made.

The effect of high pressure into the damage analysis was introduced in collaboration with Liu (1994) based on the idea of "pressure softening". This assumption was founded on the knowledge that pressure melting could have a significant effect on ice response, associated with local stress concentrations between grains. For ice at $-10°C$, the pressure required to cause melting of the ice is of the order of 135 MPa for pure water (linear interpretation, Section 2.5) and 110 MPa based on Nordell's test results; the value might be closer to 100 MPa for air-saturated water. The value of 100 MPa is mentioned by Simonson et al. (1975). The maximum contact pressures found in the medium scale tests, and in ship rams, were of the order of 70 MPa. The additional damage term S_2 for high pressures as suggested in equation (6.79) was as follows:

$$S_2 = \int_0^t f_2(p) \left(\frac{s}{s_0}\right)^{q_2} d\tau. \tag{7.19}$$

Promising results were obtained that showed that a sharp drop in load was possible to obtain with the inclusion of pressure softening, so that one aspect of reality seemed to be attainable. A detailed analysis is described in Chapter 8.

7.4 Targeted Triaxial Tests

The objective of the testing programme was to model viscoelastic stress–strain behaviour under a variety of stress states, including the effects of damage, with the aim of developing reasonable inputs into continuum mechanics models of ice compression (Figure 7.1). In FE analysis, interaction with neighbouring material is of course taken into account through equilibrium and compatibility so that a flow law is required to address time-dependence, not a yield envelope. Preliminary results had indicated that recrystallization as well as microcracking were factors in the "damage" analysis. It was considered important to see if the microstructural changes corresponded to the changes in the hpz "layers" observed in the field, and to develop a viscoelastic model to apply in the analysis of the mechanics of a high-pressure zone, reflecting nonlinearity and damage based on measured data. Figure 7.3 illustrates the applied

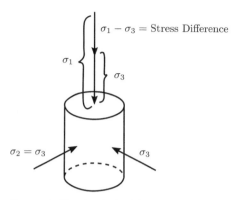

Figure 7.3 Sketch of triaxial test specimen subjected to hydrostatic pressure and axial (deviatoric) stress. From Stone et al. (1997), with permission from Cambridge University Press.

stresses to the cylindrical specimen; the stress difference applied axially is the von Mises stress, a special case being the stress in a uniaxial test; see equations (7.6) and (7.7). See also Barrette and Jordaan (2001a).

A brief note on the damage observed in field tests, as well as in small-scale laboratory indentation samples, will now be given, as it is strikingly different from the "shear faulting" found in some triaxial tests. The localization is invariably into a layer adjacent to the structure imposing the load. Shear faults as observed in some triaxial testing (including some of ours) are not seen in ice–structure interactions. It seems a reasonable deduction that the localization is a result of, and triggered by, stress concentrations. The stress concentration associated with the formation of *hpzs* is very large, associated with the transition from approximately uniform stressing in the far field to the highly concentrated high-pressure zones. The layers observed in ice crushing reflect this geometric situation of intense stress concentration, regardless of details of the crystal structure. To minimize the effects of ice structure, the triaxial testing was carried out using granular ice that could be reproduced in other test programmes to supplement the information base. Irregularities associated with crystal size and structure are treated as a separate subject.

The tests were carried out on specimens of granular, equiaxed polycrystalline ice with minimum air content. They are described in Stone et al. (1997), Melanson et al. (1999), Meglis et al. (1999) and Barrette and Jordaan (2003). The test methods and preparation of samples have been described in the papers quoted (especially Stone et al. (1997) and Melanson et al. (1999)) and are summarized in Appendix B. Figure 7.4 shows the microstructure from thin sections. The specimens were cylindrical in shape and sealed in a latex jacket before testing. Proper sealing is especially important to ensure that the confining fluid (silicone oil in the present instance) does not enter into the specimen during testing. Tests were conducted both at constant strain rate and under constant axial load applied to an already confined specimen. The confining pressure is written as $P_c (= \sigma_3)$. The axial stress difference in the triaxial test, $(\sigma_1 - \sigma_3)$,

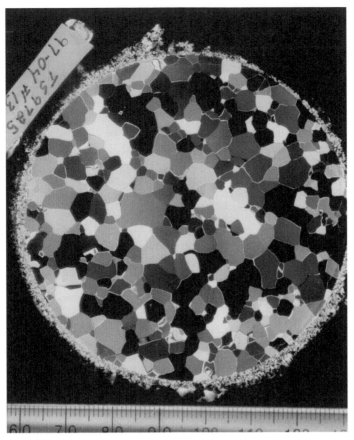

Figure 7.4 Thin section of undeformed ice viewed between crossed polarizing filters. From Meglis et al. (1999), with permission from Cambridge University Press.

is, as noted, the von Mises stress s, and we have presented values of hydrostatic pressure $p = P_c + s/3$, in addition to the confining pressure $P_c = \sigma_3$, in the following, depending on the context.

Given that considerable changes to the microstructure had been observed in the *hpz* layers in the field (Section 4.6), it was taken as being of utmost importance to track the microstructural changes in the triaxially stressed specimens to see whether comparable textures might emerge. Further, the results were intended to be used in ice constitutive modelling, as it had already been observed that the microstructural changes have a profound effect on the ice response to stress. Results are expressed in terms of true stress and true strain. The main failure modes are shown in Figure 7.5; most failures were relatively uniform, especially in the central portion of the specimen. Because of the significant lateral deformation at high strains, a substantial increase in cross-sectional area can occur. In these cases a correction was made to the engineering stress value, so as to account for the increase in area, and consequent decrease in true stress, on the basis of a constant-volume assumption, resulting in the value s_c.

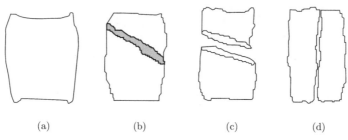

(a)   (b)   (c)   (d)

Figure 7.5 Schematic representation of macroscopic shape of deformed cylindrical specimens after triaxial loading: (a) Most specimens deform relatively uniformly; (b) strain localization, usually occurring after uniform deformation; (c) rupture generally associated with strain localization; (d) axial splitting, at very low confinement. See also Jordaan and Barrette (2014). Reproduced with permission from ASME.

A related correction was made to the strain rate (Melanson, 1998; Li, 2002; Turner, 2018), using the relationship

$$\dot{\epsilon}_c = \dot{\epsilon}_m \left(\frac{s_c}{s_m} \right)^n, \qquad (7.20)$$

where the subscripts c and m refer to "corrected" and "measured". In the case of shear stress s, the "corrected stress" is the targeted constant true stress for the test, and the "measured" value is the applied stress adjusted for the increase in specimen area. The exponent n is estimated based on measurements, with the value typically $n = 4.2$. There is another correction for damage analysis, whereby the damage that would have occurred under the higher stress s_c is taken into account.

Thin sections were generally photographed at two magnifications under three different lighting conditions. These were: (a) the thin section between crossed polarizing filters, (b) the thin section between crossed polarizing filters with side lighting, and (c) with side lighting only. Viewed between crossed polarizing filters, the grain structure is emphasized. With additional side lighting, cracks and voids can be seen in relation to the grain structure. With the thin sections illuminated only from the side, cracks, bubbles and voids can be seen more clearly.

Stone et al. (1997) conducted a detailed investigation into the effect of "damage" on the response of triaxially confined specimens. We discuss first some results of confined ice specimens tested under a confining pressure $P_c = 20$ MPa. Figure 7.6 shows strain for both an intact and previously damaged specimens. Previous damage was caused by subjecting otherwise undamaged (intact) samples to a confinement of 20 MPa and loading at a strain rate of 10^{-4} s^{-1} to a total axial strain of 2% or 4%. The damaged ice showed mainly recrystallization and possibly generation of dislocations with less microcracking than under uniaxial conditions. This was one result amongst an exhaustive set of results on damage with tests conducted under confinements of 20 MPa or less. The strains of the damaged specimen under the 7 MPa load (Figure 7.6) show the first signs of accelerating strain under stress, obtained in the present programme. Extensive microcracking and recrystallization were observed in the specimens after testing with enhanced strain rates; this was a feature of all microstructures.

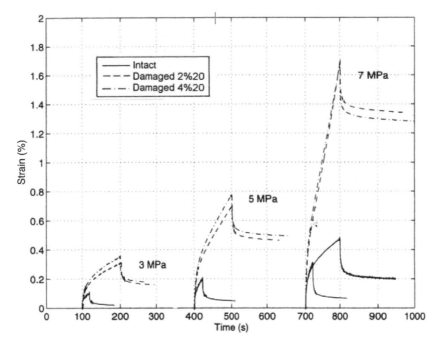

Figure 7.6 Creep response of intact and 2% and 4% triaxially damaged ice (2%20, 4%20) under 20 MPa triaxial confinement for 3, 5 and 7 MPa, 100 s pulse loads (Stone et al., 1997). With permission from Cambridge University Press.

One can conjecture that grain boundary sliding in ice under confinement would be much enhanced (see Section 7.5 regarding the effect of pressure and the presence of liquid at grain boundaries). Melanson et al. (1999) also used a confining pressure of 20 MPa in constant strain-rate tests. The results exhibited features similar to Figure 6.11, with a period of damage leading to a plateau.

Meglis et al. (1999) conducted a range of experiments, at confining pressures between 5 and 60 MPa, with constant deviatoric stresses applied as step loading on triaxially stressed specimens. The focus was to understand in a systematic manner how ice structure might break down following the scheme for different confining pressures suggested in Figure 7.1, and also to aid in developing a damage model considering time-dependence. With regard to breakdown of structure, the main results were for ice at confining pressures P_c equal to 5 and 50 MPa. The specimens were all loaded to $s = 15$ MPa and this stess was maintained until the axial strains in the specimens were 4%, 10% or 44%. At all of these stages, the specimens were removed and subjected to analysis of thin sections using crossed polarizing filters, with and without side lighting. The full set of results can be found in the paper by Meglis et al. (1999); the paper gives a detailed description of the test results and analysis with colour plates of the thin sections. We now outline the principal results.

First we consider the confining pressure of 5 MPa (with $s = 15$ MPa), conceptually modelling the case of "Sample A" in Figure 7.1 located near the edge of an *hpz*

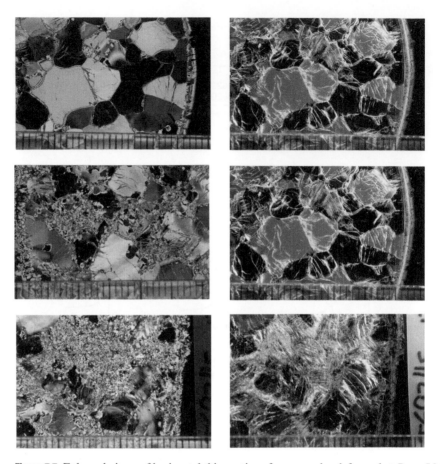

Figure 7.7 Enlarged views of horizontal thin sections from samples deformed at $P_c = 5$ MPa and $s = 15$ MPa (loading direction perpendicular to the page). Sections on the left are viewed between crossed polarizing filters alone, and, on the right, between crossed polarizing filters with additional side lighting. From Meglis et al. (1999). With permission from Cambridge University Press.

where low pressures prevail. The progression of damage was obtained for specimens deformed to strains of 4%, 10% and 44%, as already noted. The enlarged views of the thin sections are shown for these three strain levels in Figure 7.7. At the 4% strain level, one can see the microcracking commence, with some local recrystallization. Cracks are clearly seen in the views with side lighting. The cracking increases considerably together with much increased recrystallization as the imposed strain increases. Quite extraordinary images were obtained of the bands of microcracks, believed to have initiated along basal planes. We note that the cracks cannot be described as "non-interacting cracks". These are very small cracks surrounded by highly recrystallized material. At high confining pressures (50 MPa), the process consisted mainly of recrystallization starting first at grain boundaries, with very little microcracking. Figure 7.8 shows two images, one after deformation to 4% strain, the other at 44%; note that the latter is from a vertical thin section. The arrays of microcracks found in

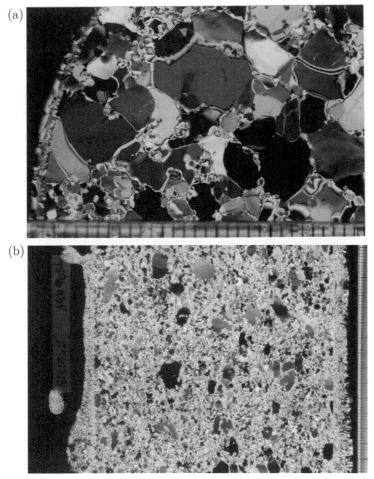

Figure 7.8 Enlarged views of thin sections from samples that were deformed at $P_c = 50$ MPa and $s = 15$ MPa to strains of 4% (a) and 44% (b) viewed between crossed polarizing filters alone. Figure (a) is a horizontal section whereas Figure (b) is taken from a vertical thin section. From Meglis et al. (1999). With permission from Cambridge University Press.

Figure 7.7 are absent. In both cases of low and high confining pressure, at high strain, "islands" of large crystals that have not been broken down "float" in the fine-grained matrix. Figure 7.9 shows the microstructure of a specimen obtained under very high confinement ($P_c = 60$ MPa) and $s = 15$ MPa. There is a gradation of damage as indicated by crystal size, and shear slip at about 45° took place near the end of the test, indicated by the protruding feature on the left edge of the specimen.

These results are very important indicators of the path forward. The textures are very similar to the "layers" found in field tests, as discussed in Section 4.7. At lower confining pressures, nearer the edge of the *hpz*, the ice is highly microfractured and recrystallized, while the ice under high pressure is extensively recrystallized. Extrusion of fine-grained products can then be explained, and the "white" appearance of the

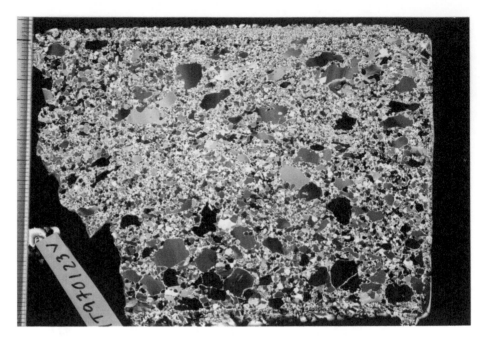

Figure 7.9 Breakdown of structure under high confining pressure $P_c = 60, s = 15$ MPa. From Meglis et al. (1999). With permission from Cambridge University Press.

layer near the edges of the layers observed in the field would correspond to microcracking and possible entrainment of air. The so-called "blue zones" observed in the field were, as observations using thin sections showed, highly recrystallized without microcracking, just as was found in the triaxial tests under high confinement. The field situation is somewhat complicated by the fact that extrusion of ice particles also takes place, but the essential features are evident.

We first show the strain measurements for tests with a constant hydrostatic pressure $p = 55$ MPa, and with a range of deviatoric stresses $s = 6, 9, 15, 18$ MPa. The results are shown in Figure 7.10. All results are characterized by a period of breakdown of structure, followed by a significantly enhanced creep rate. The breakdown corresponds to the change in microstructure as already described and is much shorter for higher shear stresses. We now look at the strain response of the ice under different confinements, having already shown the changes in microstructure. Figure 7.11 shows the strain–rate response. Some specimens showed localization and rupture as defined in Figure 7.5 (b) and (c). A close look at Figure 7.11 (note the log scale on the y-axis) will reveal that the localizations occurred when the axial strains were at about 0.4; a sudden increase in strain rates occurred, but up to that point, the low and high pressure curves (1, 2 and 3 in the figure) showed much higher rates than those at the intermediate pressures (4, 5, and 6 curves). This indicated softening due to microcracking at the low pressures being suppressed at the intermediate ones (consistent with previous results

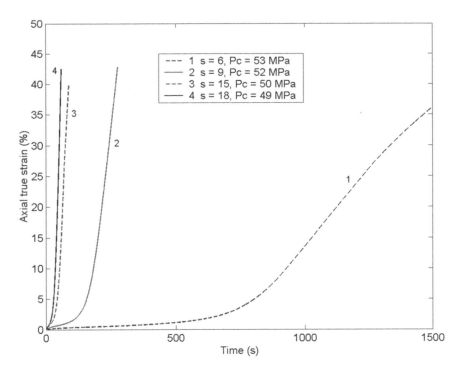

Figure 7.10 Axial true strain as a function of time, with a confinement P_c ranging from 49 MPa to 53 MPa (constant hydrostatic pressure = 55 MPa), and deviatoric stress ranging from 6 MPa to 18 MPa (after Meglis et al., 1999). With permission from Cambridge University Press.

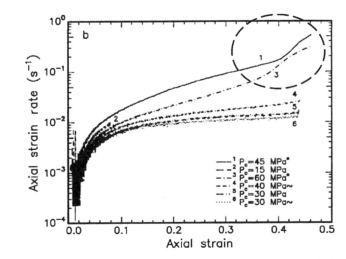

Figure 7.11 Axial true strain–rate plotted for six tests done at deviatoric stress s = 15 MPa with confinement between P_c = 15 MPa and 60 MPa. Curves marked with an asterisk are from samples which localized. Curves marked with a tilde show hard contact with ram during initial application of creep load. Circled strains exhibit runaway behaviour. After Meglis et al. (1999). With permission from Cambridge University Press.

reviewed in Section 7.2) but followed by increased rates at the higher pressures. The measured strains serve as the basis for modelling of ice as it changes its microstructure.

7.5   Barrette's Analysis

In an imaginative investigation, Paul Barrette composed a series of triaxial tests on freshwater ice subjected to a deviatoric shear stress of 15 MPa while under various levels of confinement (Barrette and Jordaan, 2003), and at a range of temperatures. A number of authors had reported an increase in activation energy of creep towards the melting point of ice. This change in activation energy near melting point has been observed only in polycrystalline ice, not in single crystals, pointing to the role of grain boundaries (Mellor and Testa, 1969; Barnes et al., 1971). It was reported that the creep rate increases as the temperature approaches melting, with a concurrent increase in activation energy from 78 kJ mol^{-1} for temperatures less than $-8°C$ to 120 kJ mol^{-1} for greater temperatures. Given our knowledge of pressure melting and its relation to temperature, Barrette conjectured that a similar phenomenon might be observed with regard to hydrostatic pressure.

The test details are as before (Appendix B). The cylindrical ice specimens were machined from polycrystalline ice made in two ways: with ultra pure water and from genuine iceberg ice. The confining pressures ranged from 10 to 65 MPa. A key part of the investigation was to vary test temperature from $-5°C$ to $-26°C$. The activation energy for the laboratory ice was determined at five distinct levels of hydrostatic pressure: 15, 35, 55, 65 and 70 MPa. The study focussed on the minimum equivalent (von Mises) strain rate and to the corresponding equivalent stress. The following expression was used:

$$\dot{\epsilon} = A_0 \sigma^n \exp\left(\frac{-Q}{RT}\right), \tag{7.21}$$

where Q is the activation energy (see also Section 2.8). Taking logarithms of equation (7.21), it can be seen that Q can be determined from the slope of plots of $\ln \dot{\epsilon}$ versus $1/T$, where T is the absolute temperature. Figure 7.12 shows a typical result. The final result regarding activation energy is shown in Figure 7.13. There resulted an increase in activation energy that mirrors that of the findings as ice temperature approaches zero noted above, from about 80 to about 120 kJ mol^{-1}. The results are also discussed in Jordaan et al. (2005b).

Further checks were made in Barrette (2014) using a range of strains (5%, 10%, 15%, 20%, 30%, and 40%) in addition to the minimum strain rate values. Similar results were found regarding activation energy. The creep rates in all cases were enhanced at the higher pressures despite the increased activation energy since there is also an increased pre-exponential term. The increase in activation energy occurred at hydrostatic pressures of about 60 MPa, somewhat less than the pressure-melting pressure of over 100 MPa for temperatures of $-10°C$. As discussed elsewhere

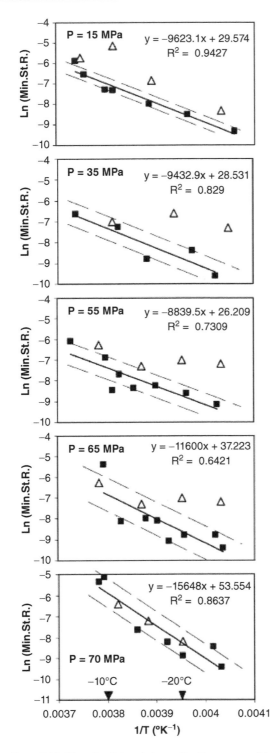

Figure 7.12 Plots for determining activation energy. Lines fitted to rectangles (granular ice from pure water). Triangles are from iceberg ice. From Barrette and Jordaan (2003). With permission from Elsevier.

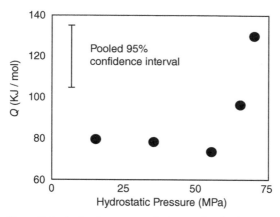

Figure 7.13 Activation energy for increasing hydrostatic pressure (Barrette and Jordaan, 2003). With permission from Elsevier.

(Barrette and Jordaan, 2003; Jordaan and Barrette, 2014), such an increase in activation energy is attributed to the presence of liquid at grain boundaries (Nye, 1991; Johari et al., 1994; Paterson, 1994, p. 86). The Barrette test series just discussed provided a large amount of additional data for analysing, supplementing that presented in Section 7.4, for use in constitutive modelling.

7.6   Localization of Damage: Runaways in Triaxial Testing

The main focus of the present work is to analyse the layer formation invariably found in ice crushing, as described extensively in past sections. Inevitably there are high stress concentrations and gradients in the ice in the vicinity of the high-pressure zones. Details of this layer have been observed in field and laboratory indentation tests; failure of the layer occurs, often with high regularity in a cyclic manner associated with a process in the layer itself, with consequent extrusion of fine ice particles; or by fractures initiated in the parent ice behind the *hpz*, which result in spalls breaking off around the edge of the *hpz*. The fractures may have some limited regularity but not of the kind found in layer failure, which are often very regular (Figure 4.13). We have found no evidence in ice–structure interaction of the "shear faults" observed in some triaxial tests. We now discuss the issue of localization with focus on laboratory triaxial tests and applicability to ice–structure interaction.

We showed in Figure 7.5 the main failure modes found in our triaxial testing. We have found in our testing that ice damage was prone to localize. Although a majority of specimens failed in a uniform manner illustrated under (a) in Figure 7.5, "runaway" strains were found in some cases, as shown in Figure 7.11. These were generally associated with distributed damage which suddenly localizes, forming shear bands after some initial widespread damage which could be quite extensive. In some cases, duplicate specimens subjected to the same conditions and test procedure as a previously tested specimen would show different results, one showing localization, the other not.

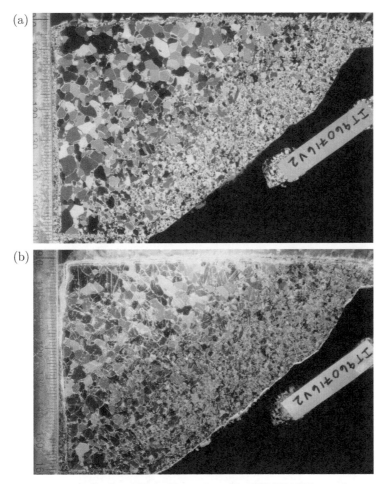

Figure 7.14 Localization followed by rupture, Test IT960716V2; specimen stressed at $P_C = 30$ MPa, $s = 30$ MPa (Meglis et al., 1999). (a) Using crossed polarizing filters alone. (b) Using crossed polarizing filters with side lighting. Reproduced with permission from Cambridge University Press.

It seems that a small local inhomogeneity in the material can, in the right circumstances, trigger localization due to the structural discontinuity. The localization does not necessarily lead to immediate failure but this does often follow a localized banding. Figure 7.9 shows a case where a small localized slip occurred, which healed after the load was taken off. An illustration of typical localization leading to fracture is shown in Figure 7.14, tested at $P_c = 30, s = 30$ MPa, after a few per cent strain. It showed extensive microcracking damage to grains throughout, and a zone of fine-grained material adjacent to the fault plane. Figure 7.15 shows a specimen of iceberg ice, tested at confining pressure of 50 MPa, in which grain refinement occurred, followed by localization and increased strain rates.

The localization generally leads to increased "runaway" strains, and eventually to rupture, but the runaway strains are important to consider in formulating

Figure 7.15 Localization in high-pressure specimen. Test IT000501, P_C = 50 MPa, s = 15 MPa, specimen of iceberg ice deformed to 31% true strain. See also Jordaan and Barrette (2014). Reproduced with permission from ASME.

stress–strain relationships. Figure 7.16 illustrates the contrast between usual and runaway behaviour. The localization appears in shear zones of increased damage and localization. Almost invariably, distributed damage occurs first, then localized faulting, rarely just faulting. Li et al. (2005) described tests on specimens of both laboratory and iceberg ice, in which the cylindrical specimen were, as before, triaxially confined at different pressure levels and subjected to deviatoric stresses. The localization was observed most often at low and high pressures, less so for pressures in between. In the case of specimens made of iceberg ice the behaviour was less pressure-dependent. In Barrette and Jordaan (2001), we reported on tests of columnar ice, with the columns

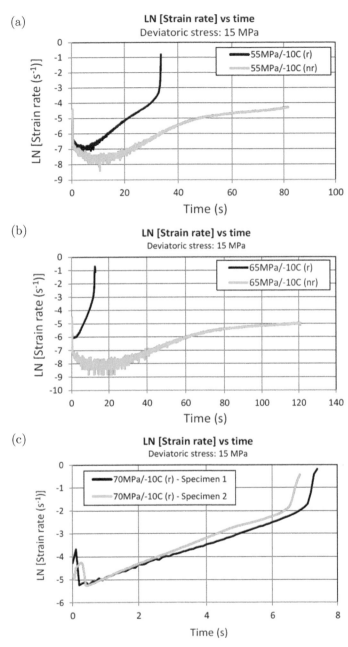

Figure 7.16 Strain rates observed in tests. (a) Tests at 55 MPa hydrostatic pressure. (b) 65 MPa hydrostatic pressure. (c) Test at hydrostatic pressure of 70 MPa. Deviatoric stress in all cases: 15 MPa. r = runaway (strain localization); nr = no localization. See also Jordaan and Barrette (2014). Reproduced with permission from ASME.

pointed in a direction normal to the direction of the application of the axial (deviatoric) stress, and with the c-axes consequently randomly oriented in a vertical plane. The tests were done with $P_c = 10, s = 15$ MPa, and with $P_c = 69, s = 30$ MPa.

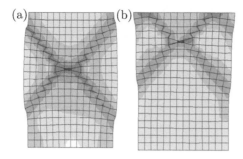

Figure 7.17 (a) Deformed mesh for the imperfection near the centre. (b) Deformed mesh for the imperfection above the centre. From Li (2002) with permission.

Both showed shear planes, with little grain refinement at the lower pressure, and considerable refinement at the higher pressure. Further research is needed to determine the role of texture (grain size distribution, irregularities) on the behaviour under confining pressure.

Needleman (1988, 1991) used time-dependent models of progressively softening materials that allow the possibility of shear localization instabilities and complete loss of strength. An imperfection in the form of a band of reduced stiffness was used to illustrate the necessary role of such imperfections in generating a shear band. In the case of ice, the banding is not planar and has finite width. In some cases, the grain refinement in adjacent material stops as energy pours into the band. Naturally these may develop along zones inclined along shear planes where macroscopic shear movements are possible, with increased deformation rate. The localization is, in our view, triggered by local inhomogeneities. An analysis by Li (2002) and Li et al. (2005) shows that this is likely to be the case, by using the constitutive relationships developed for ice (Chapter 8). This included a finite element analysis using the method described in Section 8.1. Li applied the viscoelastic damage analysis to triaxially loaded specimens under two conditions: first, restrained by the loading platens at the top and bottom, and second, with an initial flaw represented by a small area with a low elastic modulus, as compared to the parent ice. The results of the finite element analysis showed that enhanced strains occur within a band triggered by the imperfection approximately along the direction of maximum shear (Figure 7.17). The results of this study provide support for the idea that localization is triggered under triaxial conditions by discontinuities or stress concentrations rather than exhibiting a fundamental material property. The direction of research outlined represents a promising avenue for further work. It is emphasized that the formation of *hpzs* is characterized by large discontinuities and stress concentrations.

8 Damage Analysis and Layer Formation

8.1 Damage Analysis Accounting for Pressure Softening

We now outline the damage analysis which we believe leads to a reasonable mechanics-based representation of the development of the damaged layer in ice–structure interaction. The analysis is based on viscoelasticity using a Burgers model with linear springs and nonlinear dashpots. The effect of damage (microcracking and recrystallization) is taken into account by pressure-dependent state variables for each material point. These state variables are a function of past stress history. The pressure-dependence results in a U-shaped curve, with higher strain rates at low pressures reflecting microcrack damage, declining as pressure suppresses microcrack formation, and then increasing as recrystallization and local pressure melting become the dominant effects. The analyses of indentation were conducted in ABAQUS using the User Material Subroutine (UMAT) (later VUMAT in ABAQUS Explicit). The reason for this is that no standard material models exist for use in the finite element method that accurately, or even approximately, represent the material behaviour of ice.

The viscoelastic analysis based on a Burgers model with linearly elastic springs and nonlinear dashpots has been described in Section 7.1. Damage (microcracking and recrystallization) has a strong effect on the mechanical response, in particular on the strain rate, via the dashpot viscosities. The effect on the elastic properties is relatively minor, and has been discussed in Section 6.10. The introduction of pressure softening into the analysis was reported by Xiao (1997) and Jordaan et al. (1999). It was observed that, after breakdown of structure, the relationship between the logarithm of strain-rate and time was approximately linear (Figures 7.11 and 7.16; we focus on cases before or without runaway strains). We present the idealization of strain rate used in modelling in Figure 8.1. This shows the strain response under constant hydrostatic pressure and shear. The constitutive relationships follow the previous approach outlined in equations (6.79) and (7.19). The analysis, as before, exploits the linear (m_1) portion of the relationship, but attention is also given in the analyses to runaway (m_2) conditions, associated with localization.

As a first example of the methodology, we consider the strain rate in the uniaxial situation with the effect of damage, as was discussed using equations (6.77), (6.78) and (6.79). We now write the strain rate for the Maxwell dashpot in the multiaxial situation, revised from equation (7.15), as

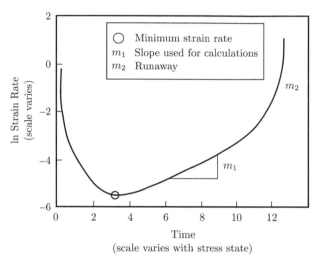

Figure 8.1 Modelling of strain–rate response (based on Barrette and Jordaan, 2002). Horizontal axis is time, typically in seconds (scale varies depending on the stress state and intensity); vertical axis is strain rate to a natural logarithmic scale which also varies depending on stress state, typically 0 to −6. With permission from IAHR.

$$\dot{e}^c_{ij} = \frac{3}{2}\dot{e}^c \frac{s_{ij}}{s} \exp\left(\beta_c S_c\right), \tag{8.1}$$

where S_c is the damage parameter (state variable) for the dashpot, and β_c is a constant to be determined from the data. We can see from this equation that, for a constant stress state,

$$\ln \dot{e}^c_{ij} \propto S_c, \tag{8.2}$$

in accordance with the approach suggested in Figure 8.1 in the linear part of the relationship. Since S_c is an integral of the stress state (representing prior stress history) with time, and if the stress state is constant, then S_c is a linear function of time, i.e. $\propto t$. This is in accord with the idealization of Figure 8.1 (m_1 slope).

To capture hardening as pressures increase, as well as softening at higher pressures, a U-shaped function of pressure emerges. This U-shaped pressure dependence is generally supported by triaxial tests, in which a suppression of microcracking occurs at the lower pressures, with a softening at the higher pressures associated with recrystallization and pressure-melting effects. Figure 8.2 shows the results of investigations of Xiao (1997) and Li (2002), as presented by Turner (2018). The damage measures are based on solutions closely related to the Schapery theory (Appendix 1), using J integral theory, originally with microcracking in mind. The nonlinear elastic theory used as a basis for the Schapery correspondence principle is the J integral, which has been described by Eshelby (1971) as the "effective force on crack tip". The energy release rate is "closely related to the force on a defect (dislocation, impurity atom, lattice vacancy and so forth)". As a result, the estimation of the energy released, implied in the damage measure, is well-based for high hydrostatic pressures as well as low ones.

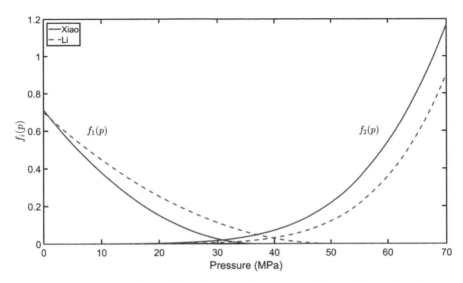

Figure 8.2 The U-shaped relationship of damage with pressure (Xiao and Li relationships). From Turner (2018) with permission.

In the example above, the Maxwell creep was enhanced as a result of damage; this could also apply to the dashpot in the Kelvin unit. In general we write:

$$S_i = \int_0^t f_i(p) \left(\frac{s}{s_0}\right)^{q_i} d\tau, \tag{8.3}$$

where pressure p is considered separately in two functions f_i; thus i takes two forms, one, $i = 1$, for low pressure, the other, $i = 2$, for high pressure. Then S_1 and S_2 are general damage parameters that encompass the effects of microcracking, recrystallization and pressure melting, with the S_1 modelling the first two of these and S_2 parameter modelling the second two. This is shown in Figure 8.2. We can write $S = S_1 + S_2$:

$$S = \int_0^t \left[f_1(p) \left(\frac{s}{s_0}\right)^{q_1} + f_2(p) \left(\frac{s}{s_0}\right)^{q_2} \right] d\tau. \tag{8.4}$$

The derivative of this is as follows:

$$\frac{dS}{dt} = f_1(p) \left(\frac{s}{s_0}\right)^{q_1} + f_2(p) \left(\frac{s}{s_0}\right)^{q_2}, \tag{8.5}$$

giving the damage rate as a function of stress expressed as a power law.

The analyses were carried out using an incremental approach. In early creep analysis, one would maintain the elastic state and let the material creep for a small interval of time, and then adjust the elastic state to ensure equilibrium and compatibility. In most of the work in the present programme, an explicit incremental scheme was followed. Following Xiao (1997), we can write the elastic relations as

$$\sigma_{ij} = C_{ijkl} \epsilon_{kl}^e. \tag{8.6}$$

For an isotropic material,

$$C_{ijkl} = \left(K - \frac{3}{2}G\right)\delta_{ij}\delta_{kl} + G\left(\delta_{ik}\delta_{jl} + \delta_{il}\delta_{jk}\right). \tag{8.7}$$

An increment in stress is given by

$$\delta\sigma_{ij} = C_{ijkl}\delta\epsilon_{kl}^{e} + \delta C_{ijkl}\epsilon_{kl}^{e}, \tag{8.8}$$

with

$$C_{ijkl} = \begin{bmatrix} K + 4G/3 & K - 2G/3 & K - 2G/3 & 0 & 0 & 0 \\ & K + 4G/3 & K - 2G/3 & 0 & 0 & 0 \\ & & K + 4G/3 & 0 & 0 & 0 \\ & & & 2G & 0 & 0 \\ & & & & 2G & 0 \\ & SYMM. & & & & 2G \end{bmatrix}. \tag{8.9}$$

The increment δC_{ijkl} can be obtained from the changes in the values of the damage parameter insofar as it affects δC via δG and δK; from equation (8.5),

$$\delta S = \left[f_1(p)\left(\frac{s}{s_0}\right)^{q_1} + f_2(p)\left(\frac{s}{s_0}\right)^{q_2}\right]\delta t. \tag{8.10}$$

Terms involving viscosity of the dashpots are also affected by damage and the elastic strain increment is expressed as

$$\delta\epsilon_{ij}^{e} = \delta\epsilon_{ij} - \delta e_{ij}^{c}, \tag{8.11}$$

where $\delta\epsilon_{ij}$ is typically imposed by boundary conditions such as indentation rate, and the creep strain is typically from equation (8.1):

$$\delta\dot{e}_{ij}^{c} = \left[\frac{3}{2}\dot{e}^{c}\frac{s_{ij}}{s}\exp(\beta_c S_c)\right]\delta t. \tag{8.12}$$

In Jordaan et al. (1999), we used the relationships:

$$f_1 = a_1\left(1 - \frac{p}{p_1}\right)^2 \tag{8.13}$$

and

$$f_2 = a_2\left(\frac{p}{p_2}\right)^r, \tag{8.14}$$

with the constants $a_1 = 0.712$, $p_1 = 37.0$ MPa, $a_2 = 0.1$, $p_2 = 42.8$ MPa, $q_1 = q_2 = 5$, $s_0 = 15$ MPa. Many of these values were also used in the subsequent analyses of Xiao, Li and Turner, described below. The value of r depends on which part of the strain–time plot is used; one might wish to model a runaway situation. The value $r = 7$ was used in the simulation now described.

The analysis was carried out using true stress and strain. Although the medium-scale tests were in the background as an objective, one can only aim at a plausible interpretation, since despite careful characterization of the failure zone, the exact spalled area, and areas corresponding to "soft" extrusion are not known with any

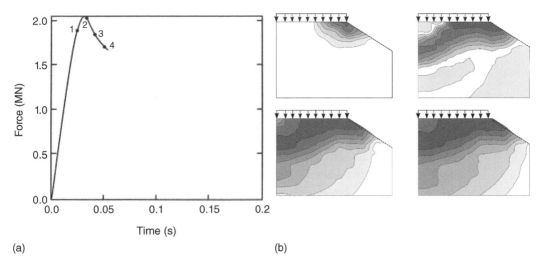

(a)   (b)

Figure 8.3 (a) Simulation results (Jordaan et al., 1999). (b) Progress of damage. The development of the layer in (b) is shown in four steps corresponding to the numbers in (a). The initial damage occurs near the edges, followed by the centre, which begins to soften; the load drop occurs when the zones join together. With permission from Springer Nature.

precision. Based on previous analyses these could be as high as 80 per cent of the nominal contact area. A high-pressure zone of circular shape, 150 mm in diameter was analysed. A small initial level of damage in the area close to the indenter was included to simulate the real (field) situation; the loading rate was set at 50 mm s^{-1} to model the observed slowing down of the indentation on the upswing in load (Meaney et al., 1996). Figure 8.3 shows the results of the analysis, in terms of the force–time curve and associated layer development. The damage state is shown by shading in the figure. The darker the shading, the greater the extent of damage, with the darkest shade corresponding to softened extruding material. This convention is followed in the subsequent figures showing damage states. The rate of loading, the peak load, the development of the layer and its shape and localization are well represented, and give a plausible interpretation of the failure of an *hpz*. This was the first solution of compressive ice failure in which a layer was found to develop as a result of the mechanics, not inserted artificially, and showing a sharp drop in load, associated with pressure softening.

Xiao (1997) made considerable contributions to our understanding of compressive ice failure and conducted several analyses. He used separate functions for the Maxwell and Kelvin units as in Section 7.1, but included damage (as in Xiao and Jordaan, 1996). Therefore two damage parameters, S_c and S_k, were proposed; one for the Maxwell, the other for the Kelvin unit. Equation (8.1) deals with the Maxwell unit; for the Kelvin unit, we have

$$\dot{e}_{ij}^{k} = \frac{3}{2}\dot{\epsilon}_{0}^{k}\left(\frac{s - G_k e^k}{\sigma_0}\right)^{m}\frac{s_{ij}}{s}\exp\left(\beta_k S_k\right), \tag{8.15}$$

with

$$\delta \dot{e}_{ij}^k = \left[\frac{3}{2} \dot{\epsilon}_0^k \left(\frac{s - G_k e^k}{\sigma_0} \right)^m \frac{s_{ij}}{s} \exp\left(\beta_k S_k\right) \right] \delta t \qquad (8.16)$$

$$= \left[\frac{3}{2} \dot{\epsilon}_0^k \left(\frac{s^k}{\sigma_0} \right)^m \frac{s_{ij}}{s} \exp\left(\beta_k S_k\right) \right] \delta t. \qquad (8.17)$$

Xiao made a modification to equation (8.4) by introducing an exponential function of shear s to replace the second power-law term of the equation. This was motivated by the work of Jonas and Müller (1969), who had proposed an expression to take into account the effect of dynamic recrystallization on the strain rate, in which the strain rate follows an exponential function of stress. Xiao's expression is as follows:

$$S(t) = \int_o^t \left[f_1(p) \left(\frac{s}{s_0} \right)^{q_1} + f_2(p) \exp\left(\frac{s}{s_0} \right) \right] d\tau. \qquad (8.18)$$

The strain increment is extended from equation (8.11) and includes delayed elastic (Kelvin) and the (small) strains associated with volume change:

$$\delta \epsilon_{ij}^e = \delta \epsilon_{ij} - \delta e_{ij}^k - \delta e_{ij}^c - \delta \epsilon_v \delta_{ij}. \qquad (8.19)$$

The state variables in the analysis were $s, s_{ij}, \sigma_v, e^c, e^k$ and S, which were updated at the end of each increment. Xiao also formulated a model with three Kelvin units to explore a complete description of delayed elasticity, which he termed a "broad-spectrum" approach. He tested the idea that the springs in the Kelvin units break down at certain points in the loading process, for instance as a result of the breaking down of structure at grain triple junctions. As a result of this breakdown, the Kelvin unit becomes a Maxwell unit, adding to the viscous flow of the material by removal of the Kelvin spring which slows the unit's flow rate. This is a good approach that is hopefully to be explored further in the future.

Xiao also explored the functions and constants in equation (8.18) that gave the best representations of the reality of layer development and observations regarding load level. He used the following functions, taking the derivative of equation (8.18) and using equations (8.13) and (8.14):

$$\frac{dS}{dt} = a_1 \left(1 - \frac{p}{p_1} \right)^2 \left(\frac{s}{s_0} \right)^{q_1} + a_2 \left(\frac{p}{p_2} \right)^r \exp\left(\frac{s}{s_0} \right), \qquad (8.20)$$

with the values of constants noted above; see Turner(2018, p. 111). Using a symmetric plane strain analysis, Xiao (1997, p. 121 and Fig 5.2) reproduced independently the results published in Jordaan et al. (1999).

Figure 8.4 shows results for $q_1 = 5$ and $r = 20$, giving a sharp drop in load and the development of a distinct, relatively thin layer of damaged material. Xiao wrote the relationship symbolically as

$$\frac{dS}{dt} \propto s^5 (1 - p)^2 + \exp(s) p^{20}. \qquad (8.21)$$

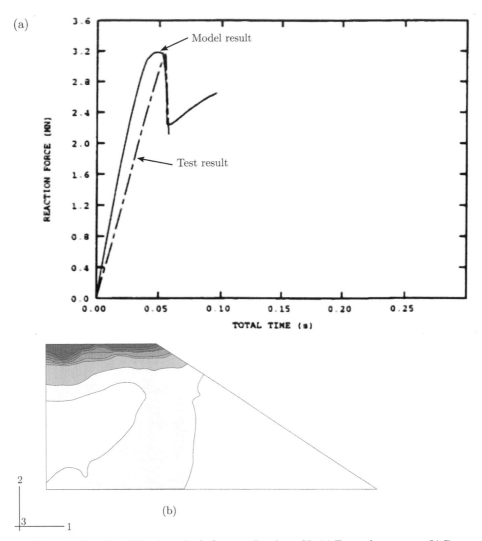

Figure 8.4 Results of Xiao's analysis for $q_1 = 5$ and $r = 20$. (a) Force-time curves. (b) Damage distribution. Based on Xiao (1997) with permission.

In the subject analysis, the inputs for the load–deformation relationship were not tailored to field conditions of test NRC 07 Hobson's Choice 1989; as shown earlier the damage constants had to include an initial damage in the ice in order to model ice in the field, and furthermore test 07 included a flexible indentor that exhibited permanent plastic deformation during the test. The latter effects are not included in the present modelling, which focusses on the ice failure. The results showed reasonable agreement with the failure load associated with the layer ultimate load.

Li (2002) used the test results of Barrette and Jordaan (2003) to evaluate the constants for the damage model thus far developed. The methodology described in Section 7.5 was used and applied to the viscoelastic behaviour of damaged ice using Barrette's test results. The merit in this approach is that the strain rates are averaged over tests at several temperatures, fitted with a regression line. The resulting calibrated relationship is illustrated in Figure 8.2, which shows Xiao relationship alongside Li's. Li presented results of an analysis, intended to simulated test NRC 07, using an axisymmetric assumption whereas Xiao had used a plane strain assumption. Quite close agreement was found in the load–deformation relationship, since he included initial softening associated with ice from the field. He also showed interesting results on the state of stress, as well as energy flux, shown in Figure 8.5. This shows the result of the softening process in layer formation, with energy pouring into the layer, and little entering the "hinterland" beyond.

Turner (2018) provided an excellent review of previous work and added original contributions. He studied layer development at various speeds of indentation, comparing the results to those obtained in small scale identation tests. This was in the first place motivated by the observations in medium scale testing (Section 4.7) that slow loading led to a permanent depression in the ice, little localized spalling, and distributed damage extending a distance into the ice (but no layer). This is mirrored by results in small scale laboratory test results (Section 8.6). The results are presented in terms of normalized velocities, defined as actual velocity divided by indentor diameter. It should be noted that the small scale indentor series was designed as a scaled

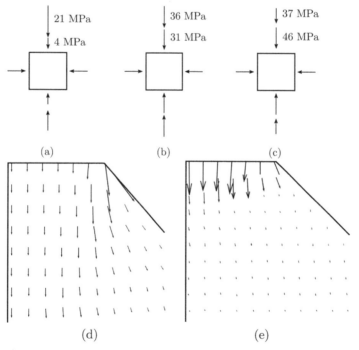

Figure 8.5 Results of Li's analysis. Top: Hydrostatic and von Mises stresses at peak load (Li, 2007): (a) near the edge of the layer, (b) between the edge and the centre, and (c) at the centre. Bottom: Flux of energy. (d) Before development of damaged layer. (e) After development of the layer. From Li (2007) with permission.

(a)

(b)

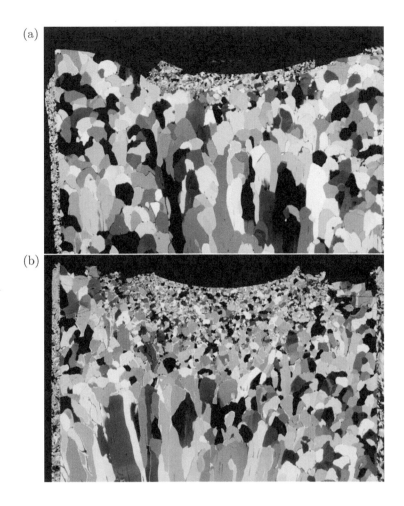

Figure 8.6 Thin sections showing damage zone for indentation at fast rate (a, T114, normalized velocity = 0.1 s^{-1}) and at slow rate (b, T130, normalized velocity = 0.01 s^{-1}), based on small scale tests. Indentor diameter was 40 mm. Photographed between crossed-polarizing filters with backlighting. See also Jordaan et al. (2016). With permission of Offshore Technology Conference.

down version of the Pond Inlet and Hobson's Choice (1989) spherical indentor series. Results are shown in Figure 8.6. The thin sections show slow and fast indentation results (40 mm diameter indentors). The "damaged" zone is clearly illustrated, with a much more extensive zone spatially in the case of the slower loading. These results are reflected in the results of the analysis shown in Figure 8.7, reported also in Jordaan et al. (2016). The results show a reasonable representation of reality, as illustrated schematically in Figure 4.12.

Turner also considered the effect of temperature, using the relation of the temperature under consideration to the pressure melting temperature. All calibration thus far has been carried out at the reference temperature, −10°C or 263°K. In Section 2.5,

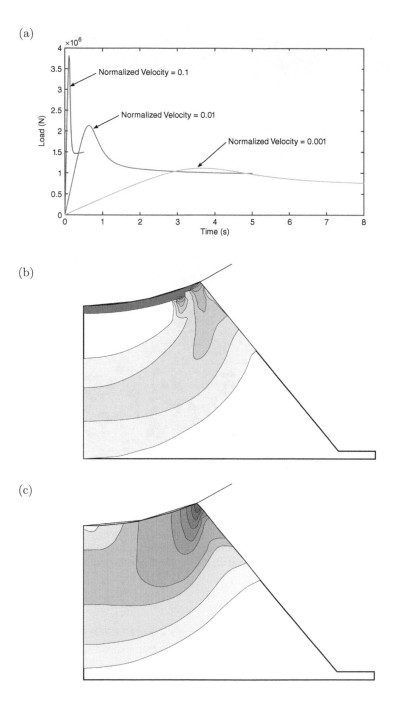

Figure 8.7 (a) Development of load with time for three normalized velocities. Forces are shown for a spherical indentor of diameter 0.2 m. (b) Distribution of damage state variable for fast loading. Highest level occurs in layer near surface. (c) Distribution of damage state variable for slow loading. Damage less severe and more distributed. See also Jordaan et al. (2016). With permission of Offshore Technology Conference.

equation (2.18), a linear relationship was derived to describe the melting temperature, with a decrease in melting temperature associated with an increase in pressure equal to $\beta°C$ per MPa, the accepted result for ice at $0°C$ being a decrease of $0.0738°C$ MPa (Hobbs, 1974). Then if we adjust the relationship $f_2(p)$ in equation (8.14) to

$$f_2\left[p + \left(\frac{(T_{ref} - T)}{\beta}\right)\right], \tag{8.22}$$

where T_{ref} is the reference temperature at which the calibration of the damage model was carried out, $-10°C$ in the present instance. In p, compression is taken as positive, as is β. For example, for $T = -5°C$, we have $f_2(p - 5/\beta)$. For negative pressures in $(p - 5/\beta)$, f_2 is zero. The right hand side of equation (8.14) remains unchanged. In this analysis, we have concentrated on the high-pressure relationship as this is the term most affected by temperature, and which dominates the failure process and load. Results are shown in Figure 8.8. At the higher temperature, the damage is more widely distributed, as expected; see also Figure 8.16. Further work could be directed to studying the effect of temperature on the low-pressure relationship $f_1(p)$, but the failure load is doubtless captured by the function $f_2(p)$. Turner also studied the effect of power-law break-down, essentially a descent into plasticity.

It is concluded that methodology has been developed, based on nonlinear viscoelasticity coupled with damage mechanics, that explains and models the formation of the viscous layer in *hpz*s in ice compressive failure, and more generally, the occurrence of microstructural change in compressive states in ice. This represents, in our opinion, the appropriate direction of research for studying the mechanics of ice failure in compression. Uniaxial tests to failure at low (1–2%) strain do not represent the actual failure conditions; triaxial test results are needed to model the failure process. Plausible results have been obtained regarding the value of failure loads, and the effects of loading rate and of temperature. Further work is needed to model the formation of the boundary between the damaged material and the parent ice, and the subsequent extrusion.

There have been some suggestions in the literature that the extremely regular fluctuations in load in indentations tests are the result of spalling. The geometry of such repeated spalls is difficult to conceive (see Jordaan, 2001), and it is here suggested that the variation is associated with "damage" failure in the layer. The analyses described have led to a convincing demonstration of layer formation and failure. The failure is essentially related to dynamic recrystallization and associated pressure softening at high hydrostatic pressures, with consequent extrusion of small ice particles. The analysis has been based on a Burgers viscoelastic model with damage, and is compatible with Schapery's MSP model. We have reviewed Kheisin's work in Section 6.4, in which a layer with linear viscoelastic properties was inserted, essentially treating the situation as a two-material problem. The present analysis provides an approach to the actual development of the layer, and its properties. The analysis showed that while elastic degradation does occur, the problem is dominated by creep effects with damage playing a major role. The material response is in reality highly nonlinear.

(a)

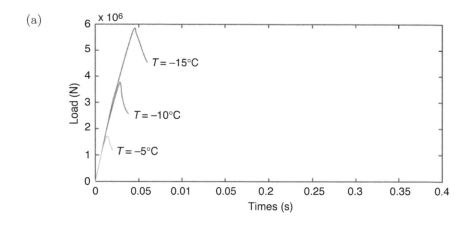

(b)   (c)

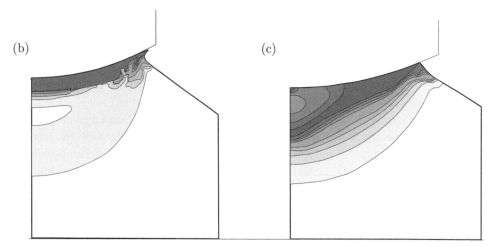

Figure 8.8 Effect of temperature on layer development. (a) Peak loads for three temperatures. Bottom: Analysis showing layer development (b) at temperature $-10°C$ and (c) at temperature $-5°C$. Based on Turner (2018) with permission.

8.2   Simplification: Incompressible Power-Law Materials

It is possible to suggest simplifications to the analysis. The inclusion of volume change might be neglected; only deviatoric strains might be included, with a constant-volume assumption. The viscoelastic response might be modelled with a single Maxwell unit once damage has occurred, noting that the flow terms become dominant as a result of damage, and the Kelvin springs (e.g. stress concentrations, triple points) will break down during the recrystallization process. The situation then becomes similar to that envisaged by Kheisin and co-workers, except that there will be considerable nonlinearity in the viscosity as it relates to stress. These ideas suggest a power-law relationship. We consider a nonlinear elastic material with stress–strain relationship

$$\frac{\epsilon}{\epsilon_0} = \left(\frac{\sigma}{\sigma_0}\right)^n,\tag{8.23}$$

where σ and ϵ stress and strain, respectively, and σ_0, ϵ_0 and n are constants. For $n = 1$, a linear elastic relationship is recovered, with σ_0/ϵ_0 equal to the elastic modulus E_0. (The present methodology could also be used to model the shear modulus G.) We consider now the potentials; first, the strain potential or strain energy function,

$$W = \int \sigma d\epsilon = \frac{(\epsilon_0 \sigma_0) n}{n + 1} \left(\frac{\epsilon}{\epsilon_0} \right)^{\frac{n+1}{n}} = \frac{(\epsilon_0 \sigma_0)}{m + 1} \left(\frac{\epsilon}{\epsilon_0} \right)^{m+1}, \qquad (8.24)$$

where $m = 1/n$, and the stress potential or complementary energy function:

$$W' = \int \epsilon d\sigma = \frac{\epsilon_0 \sigma_0}{n + 1} \left(\frac{\sigma}{\sigma_0} \right)^{n+1}. \qquad (8.25)$$

These two potentials are conjugate, related via the Legendre transformation

$$W(\epsilon) + W'(\sigma) = \sigma \epsilon, \qquad (8.26)$$

where

$$\sigma \epsilon = \epsilon_0 \sigma_0 \left(\frac{\sigma}{\sigma_0} \right)^{n+1} = \epsilon_0 \sigma_0 \left(\frac{\epsilon}{\epsilon_0} \right)^{m+1}. \qquad (8.27)$$

These are illustrated in Figure 8.9; the elastic potentials can be used in viscoelastic analysis, as described in Duva and Hutchinson (1984). We may write equation (8.23) in terms of strain rate:

$$\dot{\epsilon} = \epsilon_0 \left(\frac{\sigma}{\sigma_0} \right)^n. \qquad (8.28)$$

We note that for incompressible materials, $\epsilon_{ij} \equiv e_{ij}$, the deviatoric strain. To consider multiaxial states of stress, we write equation (8.25) in terms of the von Mises stress s:

$$W' = \frac{\epsilon_0 \sigma_0}{n + 1} \left(\frac{s}{\sigma_0} \right)^{n+1}, \qquad (8.29)$$

and obtain the strain-rate as

$$\dot{\epsilon}_{ij} = \dot{e}_{ij} = \frac{\partial W'}{\partial \sigma_{ij}} = \frac{3}{2} \left(\frac{s}{s_0} \right)^{n-1} \frac{s_{ij}}{s_0}. \qquad (8.30)$$

This approach will be exploited in Section 8.4 and in Appendix A.

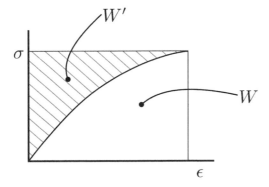

Figure 8.9 Power-law potentials.

8.3   Crushed Ice Extrusion Tests

Crushed ice is centrally involved in the failure process. It is observed to extrude as small particles during indentation and ice–structure interaction, especially when dynamic action occurs. The properties and behaviour of extruding ice were not well understood at the time of the experiments (late 1980s), and indeed nor was the creation of the small particles. The latter question has been considered in the viscoelastic–damage analyses: the small particles emerge from the high-pressure damage processes at the central portion of the *hpzs*. A special programme of testing was formulated whereby crushed ice particles were subjected to extrusion tests. The particles were squeezed between two rectangular plates 762 by 506 mm, with extrusion constrained to occur in the "long" direction. This is shown in Figure 8.10. The crushed ice layer, "loose" to start with, of 100 mm thickness and density of 0.55 g/cm^3, was squeezed between the rigid parallel plates at various speeds ranging from 2.5 mm s^{-1} to 160 mm s^{-1} at $-10°$C. The gap between the plates reduced to a few centimetres at the end of the test.

Spencer et al. (1992) reported on details of the test apparatus and presented a histogram of the sieved grain sizes. The material was loaded at constant velocity in the *y* direction. The flow of crushed ice was in the *x* direction. The channel shape of the bottom platen prevented flow in the *z* direction. Spalls were not possible because of the geometry. The top platen was fitted with eight pressure cells. Two potentiometers were mounted across upper and lower platens for displacement measurement. A closed-loop servo-controlled system was used, where averaged displacement across the platen was the feedback signal. Mean pressure was also measured at the actuator. The system as used here could apply loads close to 4 MN, with servo-control feedback.

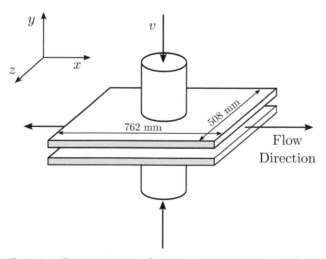

Figure 8.10 Test arrangement for experiments on extrusion of crushed ice. Flow in *z*-direction prevented so as to obtain plane–strain condition. From Spencer et al. (1992) with permission from ASME.

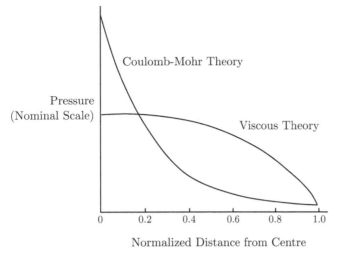

Figure 8.11 Schematic showing pressure distribution from centre to edge extrusion of material between plates based on Coulomb-Mohr and viscous theory assumptions.

Spencer et al. (1992) gave an initial assessment of the results, with further analysis being conducted by Singh et al. (1995). There were two main competing theoretical assumptions—whether Coulomb–Mohr or viscous flow theories would be relevant in analysing crushed ice mechanics. Figure 8.11 shows the pressure distributions that would be produced under these two assumptions. The results were quite clear with regard to crushed ice behaviour: at low pressures the particles behaved independently of each other and followed the Coulomb–Mohr prescription, which had been advocated by Savage et al. (1992); see also Sayed and Frederking (1992). But as the pressure increased, the particles sintered under pressure and formed a coherent mass that behaved as a nonlinear viscous material. The higher pressures are of interest in structural design where high loads are of concern in high-pressure zones. This situation was appreciated by Duthinh (1992) who presented an elegant solution for crushed ice as Coulomb–Mohr material between two circular plates.

In all cases, the ice particles were extruded relatively smoothly at first, but at the speeds of 25 and 50 mm s^{-1} dynamic action appeared. This is similar to that found in the medium scale experiments but at lower pressures. At the highest speed of 160 mm s^{-1}, a smooth curve was found. The results supported the concept of sintered ice at the centre exhibiting a convex shape of pressure distribution, with a Coulomb–Mohr "friction hill" towards the edges. This is illustrated in Figure 8.12 (a), which shows results for first (a) second half of the test at 160 mm s^{-1} and (b) a cycle in the dynamic response obtained at the rate of 25 mm s^{-1} (so-called "intermittent ductile-brittle crushing") . In both cases, a convex distribution of pressure emerges at higher pressures, indicating the behaviour of a fused mass (Figure 8.12 (b)). The results suggest a transition to viscous flow at high pressures, and sintering of the ice particles. The behaviour can be explained by the present viscoelastic theory, but not on the basis of Coulomb–Mohr theory or by spalling theories.

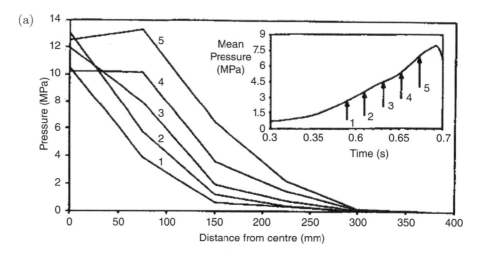

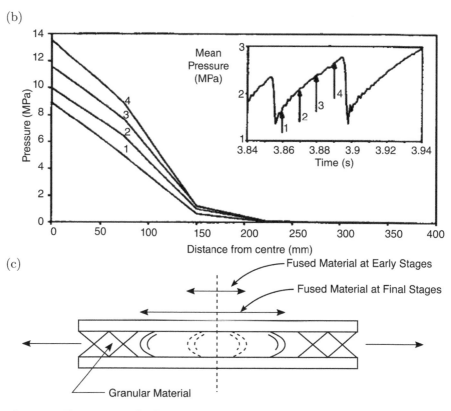

Figure 8.12 Top: Pressure distribution changing from friction-hill shape at the beginning of extrusion to a parabolic shape at higher pressures for (a) 160 mm/s (Test X996), and (b) 25 mm/s (Test X991) speed test. (c) Representation of the processes. From Spencer et al. (1992) with permission from ASME.

8.4   Singh's Analysis of Crushed Ice

The properties of crushed ice are of general interest, but particularly in the context of *hpz* behaviour. Data from medium and small scale tests, as well as the tests described in Section 8.3, show that the sharp drops and increases in load during dynamic ice–structure interaction generally occur before the entire layer has been extruded. In other words, sudden hardening of the central portion of the *hpz* occurs, related to a reversal of pressure melting during the load drop associated with softening and extrusion. This leads to solidification of the crushed ice and consequent increase in load and a repeat of the cycle. Singh (1993) undertook an investigation into the properties and behaviour of crushed ice. In Singh and Jordaan (1996) a series of triaxial tests was conducted on crushed ice. The tests were conducted at −10°C on samples prepared by crushing of ice made from deaerated distilled water. The particles ranged between sieve sizes of 1 mm and 2 mm, and were used to prepare samples of 70 mm diameter and 75 mm length. Lubricated end-platens were used to model homogeneous deformation in the sample. The initial density of samples was 550 kg m^{-3}. The axial and the radial strains, the confining pressure, the axial force, and the volume of the confining fluid movement from the pressure intensifier were measured. The test setup and instrumentation as well as details of sample preparation are given in Singh and Jordaan (1996).

Tests were conducted under hydrostatic compression, and at constant strain rate with confining pressure. Further, creep tests were conducted under constant shear stress and hydrostatic pressure. The creep tests were made possible by independent control of axial force and confining pressure. A small reduction of pressure was applied at the same time as the axial stress so that the hydrostatic pressure remained unchanged. High porosity makes crushed ice very sensitive to hydrostatic compressive stress. It was observed that, like polycrystalline ice, the deformational properties of crushed ice are nonlinear in stress and time-dependent. It was further observed that crushed ice is isotropic at high pressures. Tests of crushed ice under hydrostatic compression showed a sintering process. Figure 8.13 shows the results of creep tests.

Crushed ice was idealized as a material in which uniformly distributed cavities and grain boundaries or cracks are embedded in an isotropic matrix. In this treatment, as in the previous analyses of Section 8.1, the effect of individual microstructure is averaged and the medium is treated as homogeneous at the macroscopic level. Two state variables were identified for analysis of crushed ice, porosity and damage. Strong coupling between shear and hydrostatic responses was found. The first theory used was based on MSP using correspondence principles. Following the approach summarized in Section 8.2, Singh used the potential of equation (8.29) to define strain rather than strain-rate as

$$e_{ij} = \epsilon_{ij} = \frac{\partial W'}{\partial \sigma_{ij}} = \frac{3}{2}\left(\frac{s}{\sigma_0}\right)^{n-1}\frac{s_{ij}}{\sigma_0}. \tag{8.31}$$

This was used in Schapery's MSP model as

$$\epsilon_{ij} = E_R \int_0^t D(t-\tau)\frac{d}{dt}\left(\frac{\partial W'}{\partial \sigma_{ij}}\right)d\tau. \tag{8.32}$$

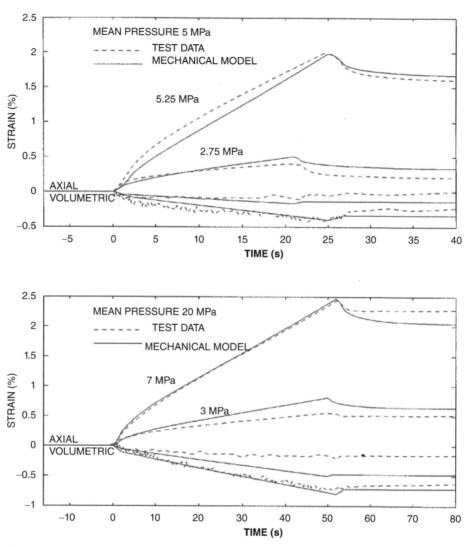

Figure 8.13 Creep and relaxation response of crushed ice at −10°C. The applied shear stresses are indicated on the curves. From Singh and Jordaan (1996) with permission from Springer Nature.

In this case "pseudostrain" or "effective elastic strain" is given by

$$\varrho_{ij} = \frac{\partial W'}{\partial \sigma_{ij}}, \tag{8.33}$$

see equation (6.98). The second viscoelastic theory is based on a mechanical model with nonlinear elements in a Burgers unit.

The analysis of the effects of microstructure on elastic properties was based on Kachanov et al. (1984) using elastic potentials, and on Budiansky et al. (1982) for analysis of dilute concentrations of spherical cavities in power-law creeping material, and finally on Duva and Hutchinson (1984) for potential functions based on this solution. The cavities in the material are spherical, isolated and noninteracting. The matrix is isotropic and incompressible. The cavities cause the potential of the system to be increased. The total potential for a material with volume fraction of cavities c is then given by

$$W_T' = W' + cW_V', \tag{8.34}$$

where c is porosity and W_V' denotes the change in complementary energy resulting from the cavities. An approximate expression from Budiansky et al. and Duva and Hutchinson is as follows:

$$W_V' = \epsilon_0\sigma_0 \left(\frac{s}{\sigma_0}\right)^{n+1} f(\alpha,n), \tag{8.35}$$

where $\alpha = 3p/(2s)$ and $f(\alpha,n)$ is a dimensionless function. Approximations for this function were developed, and the following expression for strain in a dilutely voided system resulted:

$$\epsilon_{ij} = \frac{3}{2}\epsilon_0 \left(\frac{s}{\sigma_0}\right)^n \left[\frac{s_{ij}}{s} + c\left(\frac{|\alpha|}{n} - G\right)^n \left(-\frac{nGs_{ij}}{s} + \frac{1}{3}\,\text{sign}\,(\alpha)\,\delta_{ij}\right)\right], \tag{8.36}$$

where G is a function of n and sign (α) is the sign of α.

The results so far deal with dilute concentrations. Cocks (1989) obtained a semi-empirical solution for porous materials that covers the practical range of porosities. This solution was obtained by interpolating the results for a dilute solution and a concentrated solution. The complementary energy of the porous system was given as

$$W_T' = \frac{\epsilon_0\sigma_0\,(1-c)}{n+1} \left(\frac{\bar{s}}{\sigma_0\,(1-c)}\right)^{n+1}, \tag{8.37}$$

with

$$\bar{s} = s\left[1 + c\left(\frac{2}{3} + \frac{2n\alpha^2}{(n+1)(1\_c)}\right)\right]^{1/2}. \tag{8.38}$$

The strain can be obtained by differentiating (8.37) with respect to stress σ_{ij} to obtain

$$\epsilon_{ij} = \epsilon_0 F(\sigma_{ij},c) \tag{8.39}$$

or

$$\epsilon_{ij} = \epsilon_0 \frac{3}{2}\left(\frac{\bar{s}}{\sigma_0\,(1-c)}\right)^n \left[\frac{s_{ij}}{s^2}\bar{s} + \frac{cn}{(n+1)(1+c)}\left(\frac{p}{\bar{s}}\right)\left(\delta_{ij} - \frac{3\alpha}{s}s_{ij}\right)\right]. \tag{8.40}$$

This equation defines $F\left(\sigma_{ij},c\right)$ and shows a coupling between shear and hydrostatic responses, and a strong dependence on material porosity.

Damage was represented by a similar method to that used in equations (8.3) and (8.4), dealing with microcracking (the f_1 terms) together with the associated influence of pressure. The focus was on on power-law materials and the use of the theory of Duva and Hutchinson and of Cocks to develop viscoelastic relations, using MSP and a Burgers mechanical model. In using MSP, the effective elastic strain was taken as

$$\varrho_{ij} = \epsilon_0 F\left(\sigma_{ij},c\right) g\,(S), \tag{8.41}$$

where S is the damage measure and $g\,(\cdot)$ is the enhancement of strain rate, as in equations (6.100) and (6.102) for MSP and in equations (8.1) and (8.15) for the mechanical model. The creep for MSP took the form

$$D\,(t) = D_1\,(t)^b + D_2 t. \tag{8.42}$$

The enhancement of strain was also modelled using

$$g = \exp \lambda\,(S + S_0), \tag{8.43}$$

where S_0 is the initial damage in the specimen (unity for a virgin material). In the Burgers mechanical model, a reduced elastic modulus to reflect porosity and damage was used; all other parameters regarding elasticity and viscosity were adjusted to reflect these material attributes. Details can be found in Singh (1993) and in Singh and Jordaan (1999). Results for the mechanical model are shown in Figure 8.13, in comparison with experiment.

An analysis of sintering was also undertaken. The solution for a cylindrical cavity after Wilkinson and Ashby (1975) was used. Good agreement with experiment was found.

8.5   Thermal Matters

Interesting temperature measurements were made by Gagnon and Sinha (1991) in the medium scale testing during the 1990 Hobson's Choice field indentation experiment. This was done during Test TFR0301 (flat, rigid indentor against a pyramidal ice shape). The total load was of the order of several meganewtons, and oscillating as a result of ice-induced vibrations was found. The key result is that shown in Figure 8.14. This shows the pressure and temperature measured during part of the indentation. It is to be noted first that the ice had been subjected to load for a period of time before the readings presented in the figure, with a rise in temperature associated with the viscous flow during the indentation; see Gagnon and Sinha (1991) for details of the overall temperature increase. The pressures are relatively small, but this would be expected for the high temperatures (close to melting) in the ice during the measurement. Changes in temperature of the order of 0.7°C per load cycle were found, with a drop in temperature during the upswings in load, and an increase on the downswing. The decrease in temperature arises, in our interpretation, from the fact that the locally melting ice (under pressure) will take heat from the surrounding ice in an endothermic process.

Based on the change in temperature just quoted, we estimated the volume of melted material to be of the order of 0.5 per cent of the total volume (Jordaan et al., 1999). This is not a large volume but is likely to be concentrated at the grain boundaries. A further set of results obtained using small-scale indentors was obtained by Gagnon (1994), which also shows a temperature decrease on load increases. With regard to the heating of the ice during loading as a result of viscoelastic movements, an approximate calculation was made (Jordaan, 2001) of the increase over time (i.e. of the steady

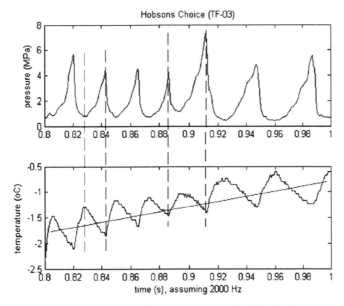

Figure 8.14 Temperature and pressure measured during indentation test Hobson's Choice 1990 Test TFR0301.

rise in Figure 8.14), averaged through the fluctuations. In the medium scale test it was estimated that about 7.5 kJ is consumed per load cycle by viscous processes and frictional effects, producing heat. A rough calculation was made on the assumption that the viscous layer is 0.02 m^3 in volume (0.2 m^2 in area and 0.1 m thick), reasonable for an *hpz* in the present circumstance; then the energy quoted would result in a warming of the ice of about 0.2°C, roughly in accordance with the measurements presented in Gagnon and Sinha (1991).

Gagnon (2002) questioned the basis of our approach regarding dynamic recrystallization, and Barrette and Jordaan (2002) provided a comprehensive reply to this position. In brief, the present model hinges on the existence of a layer of damaged ice along the ice–structure interface. Gagnon does not acknowledge this feature, and associates load drops with "spalls". The two models are discussed in some detail in Barrette and Jordaan (2002). Gagnon had suggested also that our estimate of volume of melt was 25 times too high but he based his estimate on a 14 mm crystal size. Our argument was based on the recrystallization process extensively documented, especially in the present work; the recrystallization into small grains and associated softening is the central point. Adjusting to 0.5 mm crystals, we find Gagnon's estimate 28 times too small and consequently the result is in agreement with our estimates.

The presence of a microstructurally modified layer acting as a "heat engine" might add to the knowledge of other ice–structure events such as ice-skating, and studies of change in structure of ice might also also be useful in studies of wires passing through ice (regelation).

8.6   Small Scale Tests: Synchronization

Small scale indentation tests can complement very effectively the medium scale results noted earlier, and provide new results. The first point to be made is an important one: the damage analysis described scales with linear dimension. If we consider an indentor moving at a speed u_p in the prototype (for example an indentor in the field) and at speed u_m in the model (for example a scaled-down indentor in the laboratory), the solution (equilibrium, compatibility and stress–strain equations) for the development of the layer is identical at different scales provided that the displacement rate is scaled linearly according to the scaling constant λ. In other words, the mechanics of model and prototype is identical if

$$u_p = \lambda u_m, \tag{8.44}$$

where λ is a scaling constant. One can perform a thought-experiment to support this: imagine two similar finite element meshes set up at the two scales, model and prototype. If the speed of the indentor applying load is scaled linearly, as described, the solutions are identical. The theory developed suggests that the mechanics of the *hpz* "layer" may be scaled provided that the indentation speed is adjusted linearly with linear dimension.

The first application of this approach is reported in the work of Barrette et al. (2002, 2003). The test set-up is illustrated in Figure 8.15. The steel indentor had a diameter of 20 mm and a spherical surface with a radius of curvature of 25.6 mm. This represents a scaled-down version of the spherical indentor used in the field (Frederking et al.,

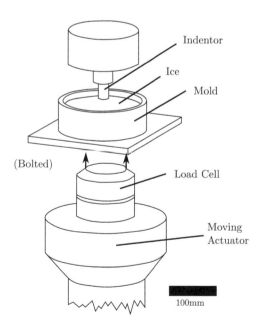

Figure 8.15 Schematic of test setup used in small-scale testing. From Barrette et al. (2002) with permission from IAHR.

1990), with a reduction factor of 50:1. The question regarding scaling just noted was posed in preparation for this series and it was decided to scale the interaction speed linearly from the medium scale tests, using a range of rates from 0.1 to 10 mm/s. Target test temperatures were −20, −10 and −2°C. The ice was seeded, generally with ice sieved between mesh sizes of 3.35 and 4.75 mm (see also Appendix B). The ice was kept confined in the mold used for freezing the specimens. This avoided premature fracture, as the main focus was on layer development.

The work covered a range of circumstances, especially temperatures and indentor speeds (scaled as noted). In general, quite remarkable agreement was found in comparisons with medium scale results, indicating that scaling of ice failure processes in compression was possible. The study did not cover details of fracture processes, which were avoided to a large extent. Regular vibrations as in full scale were observed, associated with layer failure. The load cycling was observed with peak frequencies in the order of 50–100 Hz. Fine-grained, pulverized ice was produced as a result of the interaction. Figure 8.16 displays sections from tests done at three different temperatures using an indentor 20 mm in diameter moving at 4 mm s^{-1}. For tests done at −2°C, this zone was characterized by a uniform grain refinement along most of the ice–indentor interface. For tests done at −10°C, the damage was divided into two: recrystallized areas near the centre, with a dense network of microcracks on either side. For tests done at −20°C, microcracking predominated. The features observed in these experiments, particularly those from testing done at −10°C, are very close to what is reported elsewhere from a medium-scale field testing programme with an ice surface at −14°C. This is a strong indication that the deformation mechanisms operating at both scales are similar in nature, and this reinforces the observation made above regarding scaling of the process. The report (Barrette et al. (2003)) contains a wealth of information.

Mackey et al. (2007) and Wells et al. (2011) pursued the idea of identifying fracture and damage processes in small scale testing using high-speed video, and in the second study noted, the test setup included the use of a tactile pressure sensor to measure the local pressure distribution at the indentation site (Tekscan I-scan®, see Wells et al. (2011) for details). This was motivated by the need to understand the difference between spalling and layer failure, and by the contention in parts of the literature that regular oscillation in loads were associated with "spalls". Specimens with dimensions of 20 × 20 × 10 cm were used, without restraint along the periphery (which had been characteristic of the tests reported in Barrette and Jordaan (2002)), with 20 mm indentors. Two important results from Wells et al. (2011) are given here, shown in Figures 8.17 and 8.18. These results show clearly the regularity of layer failure during crushing (Figure 8.17), where the failure corresponds to a fluctuation of pressure essentially within a defined boundary. This contrasts with the rather irregular pattern associated with spalls (Figure 8.18). The values of pressures measured in the small scale indentation tests are similar or somewhat greater than those found in the field programmes. It is also noted that the distribution of pressure across the *hpz*s is rather irregular, reflecting local fracture events and irregularities in ice structure.

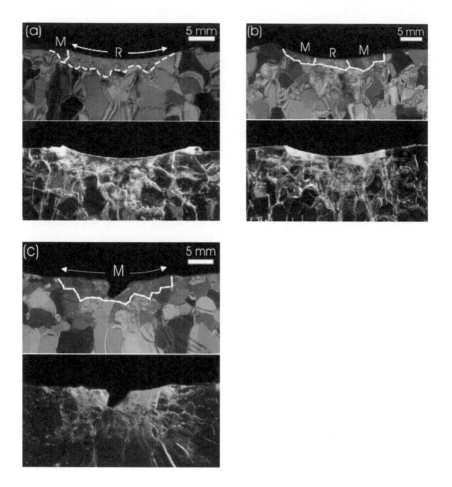

Figure 8.16 Vertical thin sections for tests carried out at (a) –2°C, (b) –10°C and (c) –20°C from Barrette et al. (2002). Each pair of images include the use of cross polarized light (upper) and plane polarized with scattered light reflected from the side (lower). Damage growth is outlined with a dotted white line in the upper photograph (R: Recrystallization, M: Microcracking). In (c), a portion of the damage zone broke off during thin sectioning. With permission from IAHR.

In the remaining series all tests were conducted with the ice confined in a steel mold, so as to concentrate on layer behaviour and development. The observations noted regarding layer development were also found by Browne (2012) and Browne et al. (2013), who also used a spherical-cap cylinder to indent the ice, again a scaled version of that used during medium-scale indentation tests at Pond Inlet and Hobson's Choice Ice Island. Diameters of 20 and 40 mm were used. Browne also tested ice at temperatures of –5°C, –10°C and –15°C. The failure of warm ice (–5°C) was dominated by ductile damage processes, while colder ice (–10°C and below) included a combination of layer failure and spalling, with increased spalling as temperature decreased. As a result, the area subjected to damage processes is greater

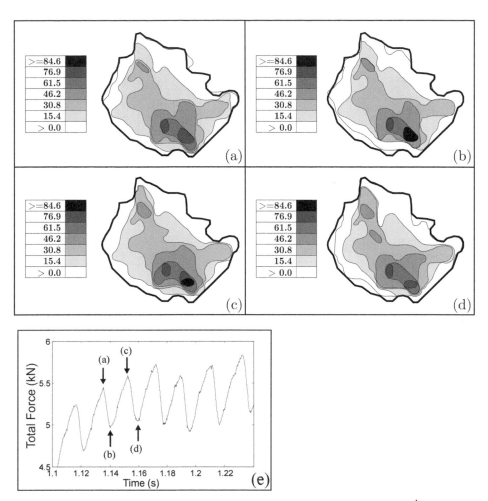

Figure 8.17 Crushing event observed during a test with an indenter speed of 2 mm s^{-1} showing layer failure: (a to d) show maps of the pressure distribution recorded during a crushing event (in MPa). The outer line in (a to d) represents the boundary of the contact area as observed at point (c). In (e) the load cycling observed in the total force (MTS load cell) time trace is shown. From Wells et al. (2011). Note peak pressures of the order of 80 MPa. From Wells et al. (2011) with permission from Elsevier.

for warm ductile ice than for colder brittle ice. During periods of cyclic loading, crushing failure is the mechanism associated with each load drop. Localized spalling can also cause a sudden decrease in load but occurs intermittently, occasionally with some limited regularity, but tends to disrupt the cyclic loading produced by crushing failure. Browne's tests also included a consideration of elastic structural feedback between the structure and the ice. Figure 4.18 showed possible structural response in terms of deflection, and these aspects were explored. Three flexible beams with varying stiffness were constructed and included in the test set-up to model the elastic structural feedback into and out of the ice specimen. Energy is stored in the deflecting structure and released upon failure of the ice. Figure 8.19 shows results at two temperatures;

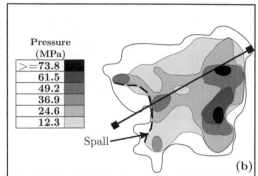

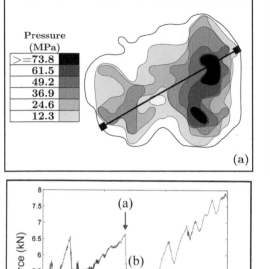

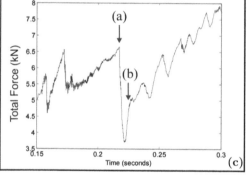

Figure 8.18 Sample spalling event showing a map of the pressure distribution (a) prior to the release of a spall ($t = 0.218$ s) and (b) after the spalling event ($t = 0.226$ s). The dashed line in (b) represents the boundary of the spall, (c) time trace of the total MTS load cell force. From Wells et al. (2011). Peak pressures greater than 70 MPa. From Wells et al. (2011) with permission from Elsevier.

although the loads are similar, the ice response is quite different, with a sawtooth response at the lower temperature of −15°C.

The results of tests with flexible indentors were promising, and a further series was conducted (O'Rourke et al. 2015, 2016a,b), incorporating a revised apparatus. Again, sawtooth loading was found in some tests, as well as lock-in, which is shown in Figure 8.20. The lock-in frequency was 204 Hz, which is 99% of the free-vibration frequency. Lock-in was found to occur during compressive ice failure when the sawtooth loading frequency approaches the natural frequency of the structure. In the lock-in condition shown in Figure 8.20, the indentor penetration is on average about 0.06 mm per cycle—significantly less than the final layer thickness. The amplitude of structural oscillation is also very small, about 0.01 mm. The indications are that this vibration is largely within the layer. In contrast, the larger-amplitude sawtooth cycling toward the end of this test reveals an average penetration per cycle of about 0.3 mm. It was hypothesized in O'Rourke et al. (2016a) that the lock-in condition may constitute vibration within the layer, whereas sawtooth loading involves periodic increases in layer thickness. As in the preceding test series, thin sections were taken at the end of each test. In all cases, layers of microstructurally modified material were found. Again,

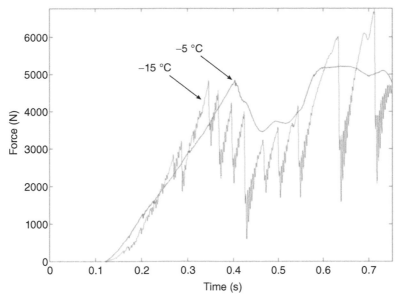

Figure 8.19 Total force versus time for tests (20 mm diameter indentor) at two temperatures (−5°C and −15°C) and an indentation speed of 4 mm s^{-1} using a compliant loading system. Note sawtooth response at the lower temperature. From Browne et al. (2013) with permission from Elsevier.

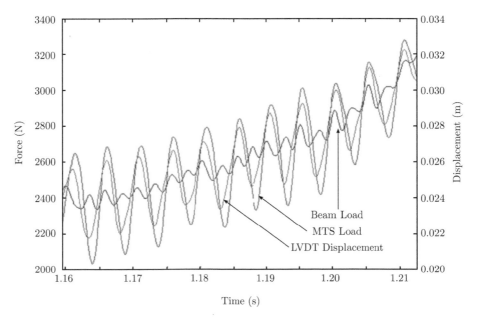

Figure 8.20 Example of lock-in during indentation at 13 mm/s using flexible beam, test T75_1C_13_10_C4. The MTS load cell applies load to the stiff beam, which supports the compliant beam at the ends. The beam load is measured between the compliant beam and the ice specimen; the LVDT measures the deflection of the flexible beam. Lock-in frequency ~204 Hz. The largest oscillations are the MTS load values, the second largest are the LVDT displacements and the smallest, the beam (indentor) load. Note that indentor load is oscillating at twice the frequency (408 Hz) of the MTS load. See O'Rourke et al. (2016b). With permission from Elsevier.

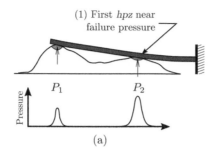

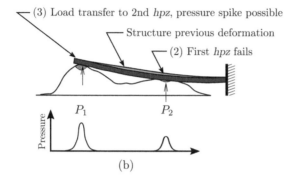

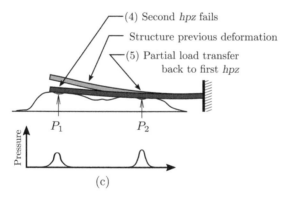

Figure 8.21 Synchronization in an ice mass failing against a cantilever structure. (a) Failure at P_2, (b) structure rebounds and load increases at P_2, (c) causing failure there. Notes (1) to (5) on figure describe sequence of events. From O'Rourke et al. (2016b) with permission from Elsevier.

as also found by Wells and by Browne, the distinction between layer and spall failure is evident. It was found that the damaged layer is thicker for the slower tests. For all tests, comparison of thin sections with indentor penetration shows that the damage layer thickness is generally much greater than the average penetration during a saw-tooth load cycle. It is interesting therefore to calculate the distance moved during each cycle in the medium scale tests for comparison with the layer thickness. For a medium scale tests in the field, typically at 10 cm s^{-1}, the layer thickness might be 10 cm, and the distance for each load cycle would be 10/50 = 0.2 cm for a 50 Hz vibration.

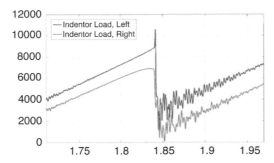

Figure 8.22 Close-up of load spikes during cyclic sawtooth loading and in-phase synchronization, test T089_2A_V04_T10_C5 with flexible beam. Horizontal axis is time (s), vertical axis is force in N. Note that the initial failure of the right hand indentor (lower curve) triggers a significant load increase in the left hand indentor, shown in the upper curve, followed by failure under the left hand indentor. From O'Rourke et al. (2016b) with permission from Elsevier.

The apparatus was also used to investigate possible synchronization of *hpz*s (O'Rourke et al., 2016b), in an effort to understand the considerable vibrations that had occurred in large structures such as the Molikpaq. In that case, the dimension of an *hpz* might be of the order of a metre in the case of thick ice, and the question is raised: how could a steel structure of size 90 m begin to vibrate under the action of *hpz*s very much smaller than the structure dimensions? The hypothesis that the vibrations might result from synchronization of *hpz*s was put forward in Jordaan and Singh (1994). The concept is illustrated in Figure 8.21 based on O'Rourke et al. (2016b), in which the failure of one *hpz* leads to the failure of another via the structural interaction. This was termed "in-phase synchronization". O'Rourke also added a second case of "anti-phase synchronization". In the work reported in O'Rourke et al. (2016b), the test setup was similar to that used before, except that there were two indentors (20 mm diameter) separated by a distance of 178 mm centre-to-centre, attached to the same beam. It was found that failure of one *hpz* very rapidly triggers failure in the other; this is illustrated with a magnified plot in Figure 8.22.

With regard to synchronization theory, it became evident that there were other events of a similar kind. The phrase "odd kind of sympathy" was used by Dutch physicist Christiaan Huygens (1629–95) in a letter to the Royal Society of London explaining the tendency of two pendulums to synchronize, when mounted together on the same wooden beam. Huygens, the inventor of the pendulum clock while trying to solve the "longitude problem", observed the effect while lying in bed. Two pendulum clocks, mounted at the same time, always ended by swinging in exactly opposite directions, regardless of their starting point. Figure 8.23 shows Huygens' sketch of the clocks suspended from a beam. He hypothesized that weak coupling of the two clocks through deformations of the beam was the cause of this anti-phase synchronization. This reasoning is similar to ours with regard to synchronization of *hpz*s; see also Jordaan and Barrette (2014). The vibration can be thought of as a self-sustained oscillation (Pikovsky et al. (2001)) in a system comprising the moving indentor which

Figure 8.23 Original drawing by Huygens of his clocks.

supplies an internal source of energy. This energy is dissipated within the layer of microstructurally modified material. The regular behaviour will often be modified or interrupted by changes in geometry and spalls. See also Palmer and Bjerkås (2013) for an interesting discussion of the connection to synchronization theory.

8.7   Scaling in Ice Mechanics

Ice tank tests are often used as an aid in the design of vessels and offshore structures for use in ice-covered waters. The use of scale models in these test basins has focussed on resistance to ship motion and on flexural failure of the ice. This has been reasonably well addressed. The properties of the model ice have often been modified to permit scaling of flexural strength as well as elastic modulus to achieve appropriate behaviour, in effect modelling icebreaking by flexural failure of the ice sheet. Useful results are obtained, but questions remain regarding the accuracy of forces and pressures so determined and scaled to full size. By way of contrast, if scaling is done in an appropriate way in problems involving fluid flow, accurate representations in certain domains can be obtained. This is based on knowledge of fluid mechanics in some detail (Navier–Stokes equations). The situation with regard to ice is not so encouraging; the mechanics are also complex but with fracture processes intervening often in a random manner.

In fluid mechanics it is possible to obtain dynamic similitude using dimensionless numbers. For example, the Reynolds number is a dimensionless number (uL/v) where u is velocity of the object with relation to the fluid, L is a characteristic length (for instance, the diameter of a pipe in studies of flow through a pipe) and v is kinematic viscosity. It represents the ratio of inertial forces to viscous forces and describes the relative importance of these two types of forces for given flow conditions. Reynolds numbers are also used to characterize laminar and turbulent flow. Laminar flow occurs at low Reynolds numbers, where viscous forces are dominant, with smooth fluid motion, not varying in time. At high Reynolds numbers, flow is turbulent and is dominated by inertial forces, which produce chaotic motions, eddies, vortices and flow instabilities. Reynolds numbers can be used to obtain dynamic similitude between model and prototype. There is an analogy to ice behaviour: at low interaction speeds ice creeps in a relatively uniform manner whereas at higher speeds, localization of damage and fracture occur in the ice, a much more random and chaotic process with regard to fracture and location and size of *hpz*s. At present, no number exists clearly to distinguish these behaviours in ice mechanics.

Michel (1978) has reviewed applications to ice. Geometric similarity follows from linear scaling of the model from the prototype. Generally Froude scaling is advocated in applications. This is generally combined with modelling of flexural strength of the ice sheet. The Froude number Fr for a vessel's resistance in water is based on the ratio of speed to waterline length of the vessel, as defined originally by William Froude in 1868. The greater the Froude number, the greater the resistance. Palmer (2008) shows the derivation of the more general form of number $\left(u/\sqrt{gL}\right)$ as a dimensionless group, and outlines its application. Here, the symbols are as before, with g representing the acceleration due to gravity. The Froude scaling deals with inertial effects in the fluid and wave-making associated with gravity and inertia of the fluid concerned. In order to include viscous effects, Reynolds number has to be introduced; Palmer describes the difficulties that arise in simultaneous modelling of inertial and viscous effects, requiring a liquid with a very high density and very low viscosity. This turns out not to be possible, so that generally modelling is done with Froude similarity using water to study inertial effects, with viscous effects dealt with separately. In some situations where clearing of ice is important, inertial effects may be a factor, but the inertial effects in ice crushing are negligible. Strong reliance on the Froude number is therefore not justified for many ice failure processes. Palmer and Dempsey (2009) have also correctly emphasized this aspect as well as the importance of fracture.

Since fracture is an important aspect of ice behaviour, it is necessary to include this in our consideration. It is the cause of the scale effect in ice–structure interaction. The reduction of global pressures with increasing area is a well-known observation for brittle materials as described in Section 5.4. The relationship to fracture toughness was explored by Atkins and Caddell (1974) and by Atkins and Mai (1985) who proposed an Ice Number in scaling. The number was based on inertia and fracture toughness:

$$I_n = \frac{u^2 \rho \sqrt{L}}{K_c}, \tag{8.45}$$

where ρ is density and K_c the fracture toughness. Apart from the consideration of inertia, the important question is: what length to use? Crack length of some kind is the obvious answer. One can consider uniform flaw structures, those that scale geometrically, and finally probabilistic distributions of flaws. The first does not lead to a scaling law, whereas the second is deemed to embody rather arbitrary assumptions: should an ice field against a 100 m diameter structure necessarily contain a flaw 1 m long if a 1 m specimen contains a flaw 1 cm in length? This does not give the observed decrease of approximately $\propto a^{-1/2}$ (Jordaan and Pond, 2001), but rather $\propto a^{-1/4}$.

Ice contains a random array of flaws. Dieter (1986) describes the physical situation associated with the scale effect. The analysis of brittle materials such as glass, ceramics or ice shows variability of results requiring statistical analysis. As a result, probabilistic modelling is used. Early manifestations of the scale effect were found in testing different diameters of glass rods (the well known tests of Griffith, discussed in Jordaan (2005)). The strength was found to decrease with diameter; further, if the tested piece was broken into fragments and the fragments tested, these would in turn yield increased strengths. Measurements of ice forces are surprisingly random. Naturally occurring materials contain flaws such as cracks and inhomogeneities such as grain boundaries. A natural way to consider this situation is to use "weakest-link" theories (Weibull, 1939, 1951; Bolotin, 1969). Elements containing flaws of different size are included in the analysis and failure occurs when the weakest element fails. The results show statistical variation, and the scale effect results from the fact that larger volumes of material have a larger probability of containing a critical flaw (reviewed in Jordaan, 2005).

It is therefore suggested that the Weibull theory and modulus might represent a possible direction of analysis for comparing an ice sheet in the laboratory and in the field (model and prototype). To pursue the theory further, we consider a set of links in a chain. Let T_i be the strength of the ith element or link, and let the strengths of the elements be considered to be independent and identically distributed (*iid*) with distribution function $F_T(t)$, for each of the $i = 1, \ldots, n$ links of the chain. The chain fails when the weakest link fails. We denote this value as $\{R, r\}$. Thus $R = \min(T_1, T_2, \ldots T_n)$. For *iid* random quantities, the distribution of the minimum is

$$F_R(r) = 1 - [1 - F_T(r)]^n . \tag{8.46}$$

This can be written as

$$F_R(r) = 1 - \exp\{n \ln[1 - F_T(r)]\}. \tag{8.47}$$

If the solid or structure under load can be divided into n elements of volume v_0, with total volume $v = nv_0$, then

$$F_R(r) = 1 - \exp\left\{\frac{v}{v_0} \ln[1 - F_T(r)]\right\}. \tag{8.48}$$

Weibull suggested that a material function $m(r)$ be used to represent the term in curly brackets { } in this equation, and in particular the power-law expression

$$m(r) = \left(\frac{r - r_0}{r_1}\right)^\alpha, \tag{8.49}$$

where r_0, r_1 and α are constants. Thus the expression for the distribution of equation (8.48) becomes

$$F_R(r) = 1 - \exp\left[-\frac{v}{v_0}\left(\frac{r - r_0}{r_1}\right)^\alpha\right]. \tag{8.50}$$

It turns out that this equation is identical to the type III extreme value distribution, so that we can interpret the Weibull distribution as the asymptotic distribution of the minimum of a set of random strengths with a lower value r_0. We shall assume that $r_0 = 0$ in the following.

The analysis can be extended to the case where the stress state is general and inhomogeneous. In this case Weibull suggested that a homogeneous stress state be assumed in a small volume at a point with coordinates represented by x. Failure criteria have to be specified, for example maximum elastic strain (Bolotin (1969)), and several others have been used. After some manipulation (see Jordaan, 2005) we arrive at

$$F_R(r) = 1 - \exp\left[-\frac{v_\star}{v_0}\left(\frac{r}{r_1}\right)^\alpha\right],$$ (8.51)

where $v_\star$ is a "reduced volume". Scale effects arise naturally from the Weibull theory. We may compare the mean of the strength $\langle R \rangle$ of two volumes v_1 and v_2 as

$$\frac{\langle R \rangle_1}{\langle R \rangle_2} = \left(\frac{v_2}{v_1}\right)^{1/\alpha}.$$ (8.52)

The weakest link hypothesis applies naturally to brittle material under tension, since increasing load will result in failure of the specimen. Failure in compression, on the other hand, implies the occurrence of spalls, splitting and in general, events of varying importance with regard to reduction in nominal stresses. A survey of literature has shown values of α for ice to range between 1.7 and 5.85 with a strong grouping between 2 and 3. Maes (1992) has also proposed a Poisson field of flaws in ice, together with a criterion based on Griffith energy release rate, constituting a promising approach.

Bazant and Planas (1998) criticized the Weibull approach with regard to sharp cracks. They inserted the equations of the stress field around the crack tip, together with singularity, into the Weibull equation and found that the integral diverged. This seemed to be taking the approach beyond an appropriate starting point. Furthermore, the crack-tip singularity is a weakness in fracture mechanics theory—infinite stresses in a body that cannot sustain such stresses, an inadequacy that was overcome by Barenblatt (1962) and others. We do not see the result (divergence of the integral under the conditions noted) as invalidating the Weibull approach, which must of course be treated as a theory with some approximation.

With regard to damage and localization, Figure 8.24 from Jordaan et al. (2012) shows the damaged layer at two very different scales. The formation of *hpz*s does show some irregularity in shape and other details, but the essential features scale quite well, as defined by equation (8.44). The mechanics of damaged layers at two scales has been discussed, with strong similarities in behaviour. This is an area where a real scale effect has been obtained theoretically and has also been observed in experiments. This could be exploited in ice tank work where compressive failure takes place. It is the practice in some laboratories to measure local pressure on model structures; it is important also to compare the local pressure and *hpz* distribution in the laboratory and in the field to determine whether the scaling follows the mechanics outlined above.

A final point should be made regarding the rate of loading. Li et al. (2004) described a set of laboratory tests on indentors of 4 different sizes (10, 20, 40 and 100 mm

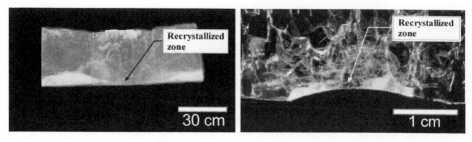

Figure 8.24 Comparison of *hpz* layers at two very different scales. From Jordaan et al. (2012) with permission from ASME.

diameter), together with a series of test results from medium scale indentor tests with indentor diameter of 1000 mm. See also Barrette et al. (2003) for details, and Jordaan et al. (2009). Tests were carried out at three sets of normalized velocities (normalized with repect to indentor diameter). We shall refer to these as "slow, medium and fast", with the slow tests being indentation with little fracture, the response being mainly creep with damage processes, much as in the slow case depicted in Figure 4.12. The medium and fast rates show the behaviour closer to that depicted in the fast case of Figure 4.12. Stresses and areas were considered on a log-plot. For slow displacement rate tests, results exhibited little scale effect, as expected for classical (in this case, mainly viscoelastic) material behaviour. As the displacement rate was increased, spalling fractures resulted in the localization of contact into zones of high pressure, and a scale effect was observed. The occurrence of spalls and fractures is important in the scale effect observed at faster displacement rates. It is concluded that the scale effect may not be observed, or is small, for slower, creep-type loadings.

9 Fracture of Ice and Its Time-Dependency

9.1   Time-Dependent Fracture in Ice–Structure Interaction

The question of ice fracture was introduced in Section 2.9. There we noted that, as with other aspects of ice deformation, fracture is time-dependent, with viscoelastic movements highly involved in the fracture process. Strong evidence in the field was found during the medium scale testing that slow indentation results in a permanent "dent" in the ice with little or no spalling, whereas for faster loading a considerable amount of spalling occurred with one or possibly more high-pressure zones at or near the centre of indentation. Yet slower loading of longer duration can lead to sudden large fractures; this was described in Section 4.7. Icebreaker captains have reported that "leaning on the ice" can lead to a time-dependent fracture of the ice floe. Large floes have been observed to split into pieces after a period under load (Figure 4.8). Small scale indentation tests confirmed the results found, with different fracture patterns at different rates. These test results were investigated in some detail by Kavanagh (2018). In summary, ice exhibits quite different behaviour at different speeds of indentation. The time-dependent growth of cracks in indentation—small or large—likely arises in the tensile zone near the indentor. A further area that needs attention is the strong effect of temperature as shown Barrette et al. (2003) and Browne (2012); certainly ice is more prone to fracture at lower temperatures.

The intense localization during ice crushing that results in the load being transmitted through only about 10% of the global area has not been addressed effectively from the standpoint of fracture mechanics. Zou et al. (1996) discuss various deterministic theories, including Kendall's work. Kendall (1978) had referred the probabilistic approach as "dubious statistical arguments involving invisible flaws". He proposed a deterministic model, well known as "the double cantilever beam". This model contained several shortcomings, as outlined in Zou et al. (1996), and it is idle to think that the failure process of ice in compression can be studied by deterministic means. A start towards a probabilistic model was described in Taylor and Jordaan (2015), as suggested in Figure 2.8.

In terms of load estimation it is contended that the small fractures and resulting spalls surrounding *hpz*s are arguably the most important fracture problem needing to be addressed. The large "floe-splitting" events might eventually lead to a reduction in load, but are not seen as an important design issue. The judgement that fractures of the order of a kilometre are of the highest engineering importance is not justified, in our view. For fast loading the use of Linear Elastic Fracture Mechanics (LEFM) can provide important insights (e.g. Taylor and Jordaan, 2015). Depending on the state of stress in the material, the crack can be subjected to tensile stress, resulting in the "opening mode" of crack separation, or to shearing stresses, resulting in the "sliding" and "tearing" modes. We shall focus on the opening mode in the following.

9.2   Measurements of Fracture Toughness

Measurements of fracture toughness are generally made under plane strain conditions. Gold (1963) drew attention to the brittleness of ice in experiments using a thermal shock technique. Goodman and Tabor (1978) measured fracture toughness in a three-point bending test and using an indentor. Fracture toughness K_c was found to be of the order of 116 kPa $\sqrt{m}$ based on the three-point bend tests. Liu and Miller (1979) and Miller (1980) showed results using compact tension specimens, in which it was shown that the fracture toughness is dependent on temperature, increasing with decreasing temperature. They also found that the fracture toughness decreased with loading rate, a trend that would be expected based on consideration of viscoelastic effects, there being more energy dissipation at the lower rates, and consequently higher fracture toughness.

Urabe (1980) completed a set of 3-point bending tests using sea ice at –2°C from an inland sea in Northern Japan. The results showed that the fracture toughness was approximately constant for strain rates less than 10^{-3} s^{-1} and underwent power-law decay for strain rates above this value. Results from our research programme, using beams tested by Kavanagh (2018) in 3-point bending using granular ice, have been shown in Figure 2.10; a three-fold range of values is seen, similar to that found by Liu and Miller (1979) in tests on columnar ice. The trend regarding temperature was also found by Hamza and Muggeridge (1979) and Nixon and Schulson (1987). In the latter work, an explanation of increase in microcracking activity was suggested, which is a likely candidate for the effect of temperature. We have found a considerable increase in cracking and microcracking activity with decrease in temperature in our indentation experiments.

Timco and Frederking (1986) performed measurements in which fracture toughness values of the order of 100 –140 kPa $\sqrt{m}$ were reported; these are, as noted, very small values compared to other materials. Dempsey (1996) and Dempsey et al. (1999) measured the fracture toughness of sea ice in two major series of field experiments: one on warm freshwater ice with a size range of 1:81 (crack lengths 0.34 to 28.64 m), and the second on first year sea ice with size range 1:160 (0.5 to 80 m crack lengths). These were significant field exercises. Base-edge-notched reversed-taper and square-plate geometries were used. The geometric configurations in the test series were self-similar. The results in sea ice covered scales 0.5 to 80 m and they found corresponding values in the range about 150 to 250 kPa $\sqrt{m}$. In assessing this, we should be aware that $\sqrt{a}$ in the expression $\sigma_f \propto K_c / \sqrt{a}$ (equation 2.27) ranges up to $\sqrt{160} = 12.6$, a large factor compared to the ratio of K_c values. Schulson and Duval (2009) questioned the conclusion that a size effect was in play, since the loading rate was not studied or controlled in a systematic way. From our point of view, time-dependent effects are unavoidable and should be part of the analysis of fracture properties. Mulmule and Dempsey (1998) outlined a viscoelastic model for the failure (process) zone. This analysis was based on linear viscoelasticity and arrived at very low stresses in the process zone, of the order of 1 MPa and less. It is not clear where the boundary of the linear material is delineated or whether the linearity applies to any material within the process zone; certainly the stresses obtained are not of the order of magnitude to represent breaking of atomic bonds. Furthermore, the size of the process zone was stated by Mulmule and Dempsey (1998) to be of the order of 2 cm, whereas the analysis by Kavanagh and Jordaan (2022), presented also in Section 9.4, suggests that the size should be of the order of 1 micron. Our analysis treats the process zone as a "singularity zone" located near the crack tip. The paper of Mulmule and Dempsey treats the process zone as a long thin strip of softened material but with very low stresses, of the order of the macroscopic tensile stress of ice (see also Kavanagh and Jordaan (2022)). As Schapery (1975a) points out, the process zone is likely to be discontinuous and highly nonlinear. The question of time-dependence is taken up in the present chapter with a somewhat different model, as compared to that of Mulmule and Dempsey.

Andrews (1985), in tests on glacier ice using a radially cracked ring fractured by internal pressure, found values as low as 58 kPa $\sqrt{m}$. He also discussed the effect of grain size. The objective is to achieve a situation where the crack tip stress field is at least one grain in extent, and resulted in a recommendation that the field itself should have a size less than one-tenth of the crack length. Given the results of the present research, it is found that the process zone is very small, so that this condition should be easily fulfilled with respect to the crack length, but hard to realize with regard to grain size. Schulson and Duval (2009) quote Nixon (1988) who found from measurements using circumferentially notched, 91 mm diameter tensile bars of granular ice (grain diameters of 3.4 mm and 7.3 mm) loaded at $-10°C$ that, as long as the depth exceeded about two grain diameters, little effect of the ratio of grain size to notch depth could be detected. This result agreed with an earlier observation by Andrews (1985). With regard to averaging over grains to represent the observations as typical for a homogeneous continuum, averaging over a large number of grains is required. This can be difficult to achieve in fracture toughness testing, but can also be dealt with by testing on multiple specimens.

As a broad statement in summary, fracture toughness values in the range of 80 to 160 kPa $\sqrt{m}$ have been found, with corresponding energy release rate of 0.5 to 2.5 J m^{-2}. Loading rate (time) and temperature are important variables in the fracture of ice.

9.3   Introductory Theory and the Barenblatt Crack

We note that when discussing the mechanics of fracture, the sign convention that tension is positive is used, applying to tensile stresses and to extensive strains. We give here an outline of theory relevant to the present analysis, but applicable to materials in general. It is by no means intended as an exhaustive account of fracture mechanics;

reference is made to the books by Broek (1986), Lawn (1993) and Anderson (2005). There have been various solutions based on elasticity theory regarding the stresses around a crack. Inglis (1913) considered elliptical cracks of dimension $(2a, 2b)$, with $2a$ being the crack length, under tension σ, and found that the stress at the tip was equal to

$$\sigma = \left(1 + 2\sqrt{\frac{a}{\rho}}\right), \tag{9.1}$$

where the radius of curvature $\rho = b^2/a$. If the radius of curvature $\rho \to 0$, then $\sigma \to \infty$. This singularity is characteristic of most analyses, and does suggest why cracks might propagate. At the same time, there are limits to the stress that any real material can carry, so that realism should enter the picture.

The elastic analysis of stresses near a crack tip was furthered by the work of Westergaard, Muskhelishvili, and Williams, amongst others, using Airy stress functions and related methodology. Kavanagh (2018) provides an excellent section in his thesis in the chapter "Critical Analysis of Linear Elastic Fracture Theory", which includes a discussion of the relevant mathematical theory and gives details of the references. We consider now the through-thickness crack with sharp tip shown in Figure 9.1. The stresses near the crack tip are given by

$$\sigma_{ij} = \frac{K_I}{\sqrt{2\pi r}} f_{ij}(\theta), \tag{9.2}$$

where K_1 is the stress intensity factor. Expressions for f_{ij} are typically of the form (Westergaard solution under biaxial tension):

$$f_{22} = \cos\frac{\theta}{2}\left[1 + \sin\frac{\theta}{2}\sin\frac{3\theta}{2}\right]; \tag{9.3}$$

the expressions apply close to the crack tip. By dimensional considerations it is found that

$$K_I = \beta\sigma\sqrt{a}, \tag{9.4}$$

where σ is the (remote) applied stress, and a is half the length of the crack shown in Figure 9.1. The dimensionless factor β depends essentially on the geometry of the crack; it is equal to $\sqrt{\pi}$ for a central crack. By convention, this factor is generally not included in the value of β, so that we write

$$K_I = \beta\sigma\sqrt{\pi a}, \tag{9.5}$$

emphasizing that β is a function of the geometry of the crack. For instance, for a penny-shaped crack in an infinite domain, $\beta = 2/\pi$. Details of its value for various geometries are readily found in the literature. Anderson (2005) gives values for common cases and laboratory specimens.

The results thus far indicate a crack tip with infinite stresses (consider small r in equation (9.2)), an unrealistic assumption dealt with by Barenblatt, as we shall see. If measurements are made on a specimen of known dimensions including crack length, the value of the stress intensity factor can be obtained at the load when failure occurs.

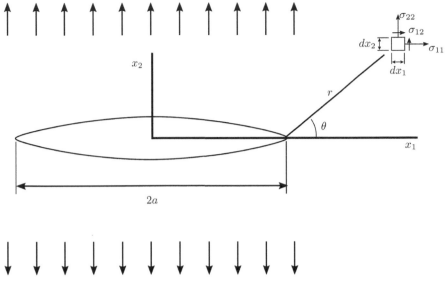

Figure 9.1 Stresses near crack tip in plate.

Common tests include tests on beams in bending, which will be used later in the present work. Then the value of K_I is described as the "critical stress intensity factor" or "fracture toughness", and denoted K_{Ic}. An important development followed the early work on crack tip stresses: consideration of the energy exchange during fracture. Griffith (1921) proposed that the loss of elastic strain energy U while a crack extends is equal to the energy required for crack growth. He considered Inglis' solution for the stress field around an elliptical crack and found that

$$\frac{dU}{da} = \frac{2\pi\sigma^2 a}{E}. \tag{9.6}$$

A feeling for this can be found by considering the elastically stressed region as illustrated in Figure 9.2, where a crack in uniaxial tension is shown. The lines of internal force illustrate areas of concentrated stress and stress-free areas. As the crack extends, the unstressed volume increases, thus releasing strain energy corresponding to strips along the sloping sides of the triangular unstressed regions in the figure. The unloaded area is approximately γa^2, taking unit thickness, with a derivative with respect to a being $2\gamma a$. The strain energy density is $\sigma^2/2E$, and the basis for equation (9.6) can be seen by solving for the energy released by a small increase in crack length.

With regard to the energy required, Griffith originally proposed that all energy released was stored as surface energy. Later it was realized that a large amount of energy (the greater part) was dissipated by plastic flow near the crack tip. In our case the dissipation would be by viscoelastic movements. The energy released in equation (9.6) is for two crack tips, so that we write

$$G = \frac{\pi\sigma^2 a}{E}, \tag{9.7}$$

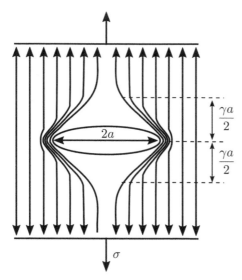

Figure 9.2 Schematic of stress distribution near a crack.

where G is the energy release rate for a single crack tip. It is noteworthy that Griffith's theory introduces explicitly the crack length a. Crack growth occurs when this is the critical energy release rate. For our mode I crack, growth occurs when

$$\sigma_c = \sqrt{\frac{EG_{Ic}}{\pi a}}. \tag{9.8}$$

Now, using equation (9.5) at crack extension with $K_I = K_{Ic}$, and $\beta = 1$, together with equation (9.8):

$$K_{Ic}^2 = E'G_{Ic}. \tag{9.9}$$

This equation has been developed for plane stress, where $E' = E$, the elastic modulus; for plane strain $E' = E/\left(1 - v^2\right)$, where v is Poisson's ratio. For ice, $v = 0.33$, $(1 - v^2) \simeq 0.9$.

Several important developments took place in the 1960s and 1970s. Barenblatt (1962) developed an analysis to deal with the singularity at the crack tip; Rice (1968) introduced the J Integral as a means of calculating the energy releases at a crack tip, strictly correct for a nonlinear elastic material, but applied by him to plastic materials, in cases without significant unstressing. Schapery (1975a,b,c) exploited the linear elastic structure of the Barenblatt theory to formulate a linear viscoelastic analysis, being careful not to apply this in the failure zone. This is illustrated in Figure 9.3, based on Schapery (1975a). The crack might be an edge crack of length a, or an internal crack as in Figure 9.1 of length $2a$. The crack tip and process zone are shown with an exaggerated vertical scale, as compared to Figure 9.1 so as to clarify the notation and assumptions. A word on nomenclature: Barenblatt considered the "forces of cohesion" across the zone, hence the term "cohesive zone". This is also termed the "process zone" and "failure zone". (A related term is "fictitious crack model", used

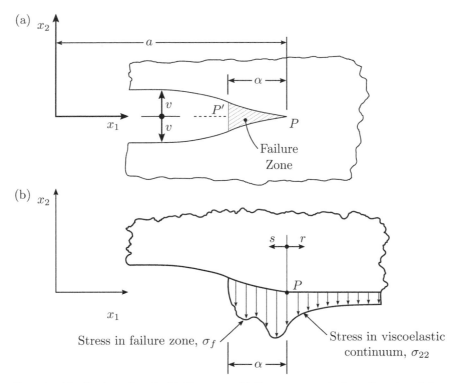

Figure 9.3 Idealization of crack. (a) Geometry. (b) Stresses on crack plane. Based on Schapery (1975a) with prmission from Springer Nature.

by researchers in the behaviour of concrete.) We shall use the term "failure zone" as coined by Schapery in the following.

Barenblatt discusses the breaking of atomic bonds and the approximation of theoretical strength of materials as being approximately equal to $E/10$, as outlined in Section 2.9. Care must be taken to avoid making arbitrary assumptions regarding behaviour in this zone. Another point to note is that in Figure 9.3 (a) the point P is described as the crack tip. Conventionally the point P' in the figure is so described, and we shall refer to this point P' as the "apparent crack tip", following Schapery. The material in the failure zone includes disintegrating and damaged material, in which the opposite crack faces are subjected to significant interatomic forces. The assumption is that it is small, very reasonable for ice with its very low energy release rate. Then it is possible to determine the fracture behaviour of ice acceptably well using an averaged model of the process zone. The smaller details of the failure zone will have little influence on the result.

The Barenblatt (1962) analysis is based on linear elasticity for material outside the failure zone; see also Lawn (1993). Schapery's extension was to convert this to linear viscoeasticity using correspondence principles. We now discuss the question of the singularity at the tip of a crack. This is exemplified by the $1/\sqrt{r}$ term in equation (9.2) for example. The analysis focusses on the shape of the crack profile (edge of

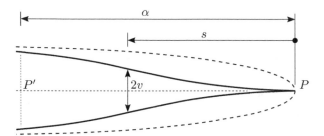

Figure 9.4 Barenblatt cusp and Irwin parabola (dashed line). Based on Lawn (1993) with permission from Cambridge University Press.

the failure zone) as shown in Figure 9.4, as a pair of cusp-shaped surfaces converging to the point P. The classical Irwin parabola is also shown. To deal with the stress field along the failure zone, Barenblatt considered a superposition of two stress fields, expressed in terms of stress intensity factors at P. The first stress field is that resulting from the external loading resulting in the usual stress intensity factor K_1. The second stress field results from the "forces of cohesion" over the crack surface. This was derived by Barenblatt using limiting values of Cauchy-type integrals. The resulting stress intensity factor is

$$K_0 = -\sqrt{\frac{2}{\pi}} \int_0^\alpha \frac{\sigma_f (s)}{\sqrt{s}} ds, \qquad (9.10)$$

based on consderation of the stress along the failure zone σ_f (Figure 9.3), necessitating the use of a minus sign (stresses pulling towards the centre of the crack). The distance s is measured from the crack tip at P to a point at x_1 within the failure zone:

$$s = a - x_1. \qquad (9.11)$$

The singularity is removed by superimposing this on the stress intensity factor associated with the external loading K_I (Westergaard or Williams solutions for example), thereby removing the singularity at the crack tip:

$$K_I + K_0 = 0, \qquad (9.12)$$

the Barenblatt condition. Therefore

$$K_I = \sqrt{\frac{2}{\pi}} \int_0^\alpha \frac{\sigma_f (s)}{\sqrt{s}} ds. \qquad (9.13)$$

The elastic displacement of the top of the crack face in the neighbourhood of the tip associated with both internal and external stress fields (K-fields) is given by the following equation (see Schapery 1975a):

$$v (s) = \frac{C_e}{2\pi} \int_0^\alpha \sigma_f (s') \left(2\sqrt{\frac{s}{s'}} - \ln \left| \frac{\sqrt{s'} + \sqrt{s}}{\sqrt{s'} - \sqrt{s}} \right| \right) ds'. \qquad (9.14)$$

The constant $C_e = 4\left(1 - v^2\right)/E$.

For small cracks in the interior of a solid, the assumption of plane strain is generally valid, given the restraint provided by surrounding material. Schapery states that the neighbourhood of the crack tip P in Figure 9.3 (Schapery crack) may be assumed to be in a state of plane strain provided: (i) the length of the failure zone α is small compared to the distance to the nearest geometric feature, (ii) α is small compared to the radius of curvature of the crack edge at P and (iii) the value of σ_{33} away from P can be neglected in comparison to that of σ_{22} near P. These conditions will generally be valid in the case that α is very small, as is the case for ice.

9.4 Schapery's Linear Viscoelastic Fracture Theory

The following analysis has been conducted using the assumption of plane strain, as just discussed. The correspondence principle in linear viscoelasticity has been outlined in Section 6.9. The principle has been extended by Graham (1968) to deal with moving boundaries, and cracks in particular. The analysis was first presented by Schapery (1975a,b,c). We start with the integral of equation (9.13) and let

$$\sigma_M = \max \left[\sigma_f(s) \right] ; f = \frac{\sigma_f(s)}{\sigma_m}. \tag{9.15}$$

Then we can write

$$K_I = \sqrt{\frac{2\alpha}{\pi}} \sigma_M I_1, \tag{9.16}$$

where

$$I_1 = \int_0^1 \frac{f}{\sqrt{\eta}} d\eta \tag{9.17}$$

with $\eta = s/\alpha$. Equation (9.15) models the distribution of σ_f across the failure zone using normalized non-dimensional stress, denoted $f(s)$.

Equation (9.16) is important for its use in estimating the size of the failure zone. Rearranging:

$$\alpha = \frac{\pi}{2} \left(\frac{K_I}{\sigma_M I_1} \right)^2. \tag{9.18}$$

Let us assume initially that the distribution of σ_f is uniform over the failure zone; then $f = 1$, and $I_1 = 2$. It then follows that

$$\alpha = \frac{\pi}{8} \left(\frac{K_1}{\sigma_M} \right)^2. \tag{9.19}$$

We can get an estimate of the magnitude of α. Taking $K_1 = 150$ kPa $\sqrt{m}$ and $\sigma_M = 10^5$ kPa, we find that $\alpha \approx 0.9$ micron. In choosing the value for σ_M, we have to consider that basal planes are prone to cleavage fracture. This is likely where fracture would initiate, depending on the orientation of the ice crystal. It is to be noted that fracture along the basal planes requires the breaking of two bonds as against four or more for other planes. The process zone is evidently quite small.

The crack opening displacement in the region of the crack tip, developed in equation (9.14), can be simplified, after some manipulations (Schapery, 1975a and Kavanagh, 2018) to the following, valid for $0 \leq s \ll \alpha$:

$$v(s) = -\frac{2C_e}{3\pi} \frac{\sigma_M}{\sqrt{\alpha}} s^{\frac{3}{2}} I_2, \tag{9.20}$$

where

$$I_2 = \int_0^1 \frac{\partial f}{\partial \eta} \frac{1}{\sqrt{\eta}} d\eta, \tag{9.21}$$

again with $\eta = s/\alpha$. This equation is also used in Section 9.5 on nonlinear analysis of crack growth.

As we increase the loading on a body, the stress intensity factor given in equation (9.16) will increase, with the length of the failure zone satisfying this equation. Therefore α will increase in proportion to K_1^2 assuming that σ_M and the shape of the failure zone (defined by f in I_1, equation (9.17)) remain similar. As the load increases, failure starts in a small area to the left of the failure zone, and the failure zone moves to the right. This will be denoted the fracture initiation time t_1. This is the time at which the process zone is fully developed, followed by continuous crack growth. It is assumed that the time interval for development of the process zone is small, as compared to the time to failure, so that we focus on the determination of the time to failure. Both aspects are dealt with by Schapery and by Kavanagh.

The growth of the crack occurs when a "column" of material of cross-sectional area dA in the failure zone fails. The criterion for this is that failure occurs when the work done on the column by the adjacent continuum outside the fracture zone reaches the value ΨdA, where Ψ is the **specific fracture energy**, or simply the fracture energy. The criterion can be expressed as

$$\Psi = \int_0^{v_m} \sigma_f dv = \int_{t_1}^{t_2} \sigma_f \frac{\partial v}{\partial t} dt, \tag{9.22}$$

where v_m is the height of the column at the point of rupture, and where $v = 0$ at t_1 and $v = v_m$ at t_2.

We now consider time-dependent deformation $v(s,t)$ using linear viscoelastic theory for the material adjacent to the failure zone. For a stationary crack, it would be relatively simple (at least in principle) to convert the elastic analyses above, but for a moving crack, the extended correspondence principle of Graham (1968) is needed. Full details are given in Schapery (1975a) and in Kavanagh (2018). The main conditions are that the crack does not decrease in size, that boundary conditions are on stress, and with regard to dependence on elastic modulus and Poisson's ratio. The latter condition specifies that dependence of the solution for elastic stress normal to the surface of the crack must be independent of E and v, while dependence of the elastic displacement of the crack face must be in the form of separate functions of s and of $\{E \text{ and } v\}$.

It is assessed that the conditions are well met for the present anaysis. In the case where the displacement of beams are specified, this can be converted to that of applied forces, by treating the applied forces as being specified, as discussed by Schapery but we have focussed on the dependence on stress in our later experimental work. Restrictions regarding Poisson's ratio are met in many cases considered, and can certainly be met if the viscoelastic material has a constant Poisson's ratio. This will be approximately true for ice. The viscoelastic theory accounts for the stress history at any point in the continuum. For a moving crack, the stress distribution changes as the crack moves. The integration is performed with x_1 being constant, and as a result s in equation (9.11) is expressed in terms of time:

$$s(x_1, \tau) = a(\tau) - x_1. \tag{9.23}$$

Given that t_1 is the crack initiation time,

$$s(x_1, t_1) = a(t_1) - x_1 = 0, \tag{9.24}$$

and therefore $a(t_1) = x_1$.

The variation of the crack opening displacement with time is important first for the calculation of fracture energy and second, for the time-dependent movement of the crack tip. In the first instance, for $0 \leq s \leq \alpha$, equation (9.14) is cast in time-dependent form using the correspondence principle:

$$v(s,t) = \frac{1}{2\pi} \int_{t_1}^{t} C_v(t-\tau) \frac{\partial}{\partial \tau} \left[\sigma_f(s') \left(2\sqrt{\frac{s}{s'}} - \ln \left| \frac{\sqrt{s'} + \sqrt{s}}{\sqrt{s'} - \sqrt{s}} \right| \right) ds' \right] d\tau, \tag{9.25}$$

and for the approximation $0 \leq s \ll \alpha$, equation (9.20) for crack opening is used to formulate the viscoelastic approach:

$$v(s,t) = -\frac{2}{3\pi} \int_{t_1}^{t} C_v(t-\tau) \frac{\partial}{\partial \tau} \left(\frac{\sigma_M}{\sqrt{\alpha}} s^{\frac{3}{2}} I_2 \right) d\tau. \tag{9.26}$$

Both of these integrals naturally involve a convolution of the time-dependent compliance $C_v(t)$, the viscoelastic creep compliance analogous to C_e in the elastic solution of equations (9.14 and 9.20).

We can substitute $\tau - t_1 = \rho, d\tau = d\rho$ and $t - t_1 = \Delta t$ in equation (9.26); thus

$$v(s,t) = -\frac{2}{3\pi} \int_{t_1}^{t} C_v(\Delta t - \rho) \frac{\partial}{\partial \rho} \left(\frac{\sigma_M}{\sqrt{\alpha}} s^{\frac{3}{2}} I_2 \right) d\rho. \tag{9.27}$$

Crack growth is measured using the distance a. We now assume that $a(t)$ can be linearized for the short time period $t - t_1$ as follows:

$$a(t) = a(t_1) + (t - t_1) \dot{a}, \tag{9.28}$$

where $\dot{a}$ is the crack tip velocity at time t_1.

The solution developed by Schapery relies upon the use of an "effective compliance" C_{eff} to solve the integrals (see Section 6.9). In equation (9.27) we take α, σ_m and I_2 as constant and find that

$$v(s, \Delta t) = \frac{2\sigma_m I_2}{3\pi \sqrt{\alpha}} s^{\frac{3}{2}} C_{eff}(\Delta t) \tag{9.29}$$

with $s(x,t) = \dot{a}\Delta t$, and

$$C_{eff}(t) = \frac{3}{2} t^{-\frac{3}{2}} \int_0^t C_v(t-\rho)\rho^{\frac{1}{2}} d\rho. \tag{9.30}$$

Schapery developed a detailed methodology for approximating C_{eff}. This is also well treated in Kavanagh (2018). The analysis relies on a power-law approximation of C_v over the influential part of the compliance:

$$C_v(t) = C_1 t^n. \tag{9.31}$$

There results

$$C_{eff}(t) = \lambda_n C_1 t^n \tag{9.32}$$

or

$$C_{eff}(t) = C_v\left(\lambda_n^{\frac{1}{n}} t\right). \tag{9.33}$$

This is very easy to use: the expression for λ_n is also easy to use. It is found in making the substitution just noted (see also Kavanagh (2018) for details):

$$\lambda_n = \left(\frac{3\sqrt{\pi}}{4(n+\frac{3}{2})}\right) \cdot \frac{\Gamma(n+1)}{\Gamma\left(n+\frac{3}{2}\right)}. \tag{9.34}$$

Here $\Gamma(\cdot)$ is the gamma function, $\Gamma(n) = \int_0^\infty t^{n-1}\exp(-t)\,dt$. From an analysis by Kavanagh (2018), an approximate value is given by

$$\lambda_n^{\frac{1}{n}} \simeq 0.3 \tag{9.35}$$

for ice.

We can now write two approximations for the crack opening displacement at time $t = s/\dot{a}$; first, based on equation (9.29) for $0 \le s \ll \alpha$:

$$v(s,t) = \frac{2\sigma_M I_2}{3\pi \sqrt{\alpha}} s^{\frac{3}{2}} C_v(\tilde{t}), \tag{9.36}$$

where

$$\tilde{t} = \lambda_n^{\frac{1}{n}} \frac{s}{\dot{a}}. \tag{9.37}$$

The second expression is the more detailed one, for $0 \le s \le \alpha$, based on equation (9.25):

$$v(s,t) = \frac{C_v(\tilde{t})}{2\pi} \frac{\partial}{\partial \tau}\left[\sigma_f(s')\left(2\sqrt{\frac{s}{s'}} - \ln\left|\frac{\sqrt{s'}+\sqrt{s}}{\sqrt{s'}-\sqrt{s}}\right|\right)ds'\right]d\tau. \tag{9.38}$$

Two periods of time in particular can be considered: the time it takes the crack to grow some distance $s < \alpha$, and the time it takes the crack to grow the length of the process zone. The latter is written $t_\alpha = \alpha/\dot{a}$.

The next quantity to be evaluated is the fracture energy Ψ, introduced in equation (9.22). We can also write the following (equivalent) expression for Ψ :

$$\Psi = \int_0^\alpha \frac{\partial v}{\partial s'} \sigma_f ds'. \tag{9.39}$$

Schapery (1975b) showed after a detailed investigation that a reasonable approximation can be expressed as

$$\Psi = \frac{1}{8} C_v (\tilde{t}_\alpha) K_1^2, \tag{9.40}$$

where

$$\tilde{t}_\alpha = \lambda^{\frac{1}{n}} \frac{\alpha}{\dot{a}} \simeq \frac{\alpha}{3\dot{a}}, \tag{9.41}$$

about one-third of the time the crack tip takes to propagate the distance α. Equation (9.40) is similar to the linear elastic solution for a material in plane strain.

We are now in a position to study crack growth. First, we use equations (9.40) and (9.41) together with equations (9.16) and (9.33) to find:

$$\frac{da}{dt} = \frac{\pi}{2} \left(\frac{C_1 K_1^2}{8\Psi} \right)^{\frac{1}{n}} \lambda^{\frac{1}{n}} \left(\frac{K_1}{\sigma_M I_1} \right)^2. \tag{9.42}$$

This is based on the compliance being represented by equation (9.31). We can improve this by modifying the compliance of equation (9.31) by adding the elastic term $C_v (0)$:

$$C_v (t) = C_0 + C_1 t^n. \tag{9.43}$$

This does not affect the validity of equation (9.40), as is shown by repeating the analysis. Next, we define the following two values, noting that $C_v (0) \leq C_v (\tilde{t}) \leq C_v (\infty)$:

$$K_{10} = \left[\frac{8\Psi}{C_v (0)} \right]^{\frac{1}{2}} ; K_{1\infty} = \left[\frac{8\Psi}{C_v (\infty)} \right]^{\frac{1}{2}}. \tag{9.44}$$

Then we can again use equations (9.40) and (9.41) to determine the crack tip velocity:

$$\frac{da}{dt} = \frac{\pi}{2} \left\{ \frac{C_1 K_1^2}{8\Psi \left[1 - \left(\frac{K_1}{K_{10}} \right)^2 \right]} \right\}^{\frac{1}{n}} \lambda^{\frac{1}{n}} \left(\frac{K_1}{\sigma_M I_1} \right)^2. \tag{9.45}$$

It is emphasized that in using equation (9.45) C_1 is obtained by fitting equation (9.43) whereas equation (9.42) is based on $C_v (t) = C_1 t^n$.

We have analysed crack growth for a fully developed failure zone in which α remains constant. In an initially unstressed material, the question of the development of the process zone becomes important in the analysis of experiments to be described. This is dealt with by Schapery, as well as discontinuous and intermittent loading. The details of fracture initiation time t_i are not pursued here but can be found in the works

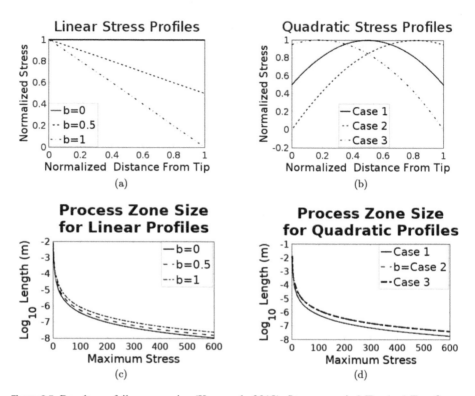

Figure 9.5 Results on failure zone size (Kavanagh, 2018). Stresses are in MPa. (a,c) Two figures show the results for a linear stress profile with b indicating the slope of the stress profile; $b = 1$ corresponds to zero stress at the apparent crack tip. (b,d) Parabolic stress profiles are shown. Reproduced with permission.

of Schapery (1975a) and were studied by Kavanagh (2018), who completed this analysis for the experiments to be described in Section 9.6. We conclude by showing the estimates of crack size by Kavanagh (2018) in Figure 9.5. The results confirm a very small failure (process) zone unless the material strength is very small.

9.5   Nonlinear Viscoelastic Fracture Theory

An approach to the problem of time-dependent fracture in material with nonlinear response to stress is outlined. We discuss first the development of the J integral (Rice, 1968). Figure 9.6 (a) shows the set-up, in which a path-independent line integral is used to determine the energy release rate from a growing crack. It is derived based on potentials and nonlinear elasticity theory. It has been applied to fracture of materials exhibiting plasticity on the assumption that there is little or no plastic unstressing so that the stress–strain curve is similar to that for a nonlinear elastic material (Rice, 1968). Schapery (1981, 1984a) used this concept to develop a generalized J integral based in part on the theory developed in the preceding section. The new direction was

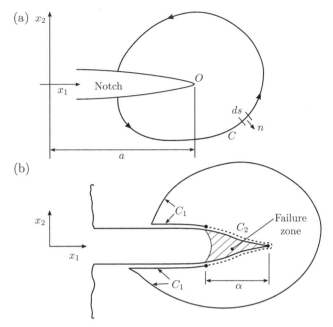

Figure 9.6 (a) J-integral analysis of energy release at crack tip. (b) Crack in nonlinear viscoelastic material as used in Schapery J-integral theory. Based on Schapery (1984a) with permission from Springer Nature.

into nonlinear viscoelastic theory. An extension to large deformations and elaboration of the theory is given in Schapery (1984b); this is not dealt with here.

The elastic-viscoelastic correspondence principle outlined in Section 6.9 and used in Section 9.4 is based on linear theory. For nonlinear materials, Schapery developed and formulated MSP theory so as to provide correspondence principles between nonlinear elastic materials and their viscoelastic counterparts, thus permitting the use of J integral theory. This is outlined in Appendix A, where the new correspondence principle linking nonlinear elastic materials to their viscoelastic counterparts is described. We use the configuration and J integral path as shown in Figure 9.6 (b), and note that the continuum may be composed of nonlinear viscoelastic material. The use of J integral theory requires that the nonlinear elasticity be expressed in terms of a potential Φ such that

$$\sigma_{ij} = \frac{d\Phi}{\partial \varrho_{ij}}, \qquad (9.46)$$

where ϱ_{ij} is taken as the "effective elastic strain" introduced in Section 6.11, and reviewed in Sections A.4 and A.5. It is the elastic strain of the nonlinear elastic material in the correspondence principle.

Implied in the approach is the requirement that the material can be well modelled by the MSP and related theory. This is believed to be amply justified in the case of ice. The J integral is a path integral defined as follows (Figure 9.6 (a) shows the established configuration, our focus is mainly on Figure 9.6 (b)), but using the nonlinear reference elastic solution in the Schapery correspondence principle:

$$J_v = \int_{C_1} \left(\Phi dx_2 - T_i \frac{\partial u_i^R}{\partial x_1} ds \right). \tag{9.47}$$

The symbol $T_i = \sigma_{ij}^R n_j$ refers to the surface tractions and u_i^R to the displacement field. The superscript R refers to the reference (nonlinear) elastic solution. The original J integral is captured if we consider a closed path in Figure 9.6 (a). The path should not enclose any singularity. The integral is zero under the following conditions (Schapery). First, plane stress, plane strain or antiplane conditions must exist inside the path; we are concerned here with plane strain. The second condition is that $\partial \Phi / \partial x_1 = \partial \Phi / \partial x_3 = 0$ on and in the closed path. The original theory exploited the fact that the faces of the crack do not have surface tractions; this was extended in Schapery's theory to consider tractions on the faces. It is to be noted that the singularity has been removed by modelling assumptions that are similar to the Barenblatt-Schapery analysis introduced in Sections 9.3 and 9.4.

We now consider the paths C_1 as in equation (9.47) and C_2 in Figure 9.6 (b), the latter adjacent to the failure zone, a departure from prior J integral theory; both paths are counter-clockwise. The paths are in the viscoelastic material to be modelled (but first considering nonlinear elasticity). We note again that there is no singularity in this theory; it follows the spirit of Barenblatt's analysis, except that the theory is based on a nonlinear elastic potential. The integral along path C_2 is denoted J_f and the condition that $\oint = 0$ along the path $C_1 + C_2$ becomes

$$J_v = J_f, \tag{9.48}$$

with J_v defined in equation (9.47) as the integral along C_1. Now the main contributions to J_f are the stresses along the crack plane. These act between initially adjacent points along the crack, now separated by the distance Δu_i:

$$J_f = \int_0^\alpha \left(\tau_{f1} \frac{\partial \Delta u_1^R}{\partial s} + \sigma_f \frac{\partial \Delta u_2^R}{\partial s} \tau_{f3} \frac{\partial \Delta u_3^R}{\partial s} \right) ds. \tag{9.49}$$

If the crack is slender, sliding movements between the points can be neglected. As a result

$$J_f = \int_0^\alpha \sigma_f \frac{\partial \Delta u_2^R}{\partial s} ds, \tag{9.50}$$

where Δu_2^R is the displacement in the x_2 direction between two originally contiguous points.

The displacement v in the linear case (Section 9.4) is one-half Δu_2; this applies to equation (9.20) and the following material, especially equations (9.36) and (9.38). We again note that for a moving crack, the stress distribution changes as the crack moves. The integration is again performed with x_1 being constant, and as a result s in equation (9.11) is expressed in terms of time; repeating equations (9.23) and (9.24):

$$s(x_1, \tau) = a(\tau) - x_1, \tag{9.51}$$

and since t_1 is the crack initiation time,

$$s(x_1, t_1) = a(t_1) - x_1 = 0, \tag{9.52}$$

and $a(t_1) = x_1$.

We now address crack propagation speed, noting that Schapery (1984a) discusses also initiation of growth. We focus as before on the time $\alpha/\dot{a}$ during which the crack propagates a distance α. First, we note that the time-dependent displacement Δu_2 is given by

$$\Delta u_2(t) = E_R \int_{t_1}^{t} D(t - \tau) \frac{\partial \Delta u_2^R}{\partial \tau} d\tau. \tag{9.53}$$

The failure criterion is similar to that used before, involving fracture energy Ψ :

$$\int_0^\alpha \sigma_f \frac{\partial \Delta u_2}{\partial s} ds = 2\Psi. \tag{9.54}$$

If σ_f and Δu_2^R can be considered to be constant for given value of s, and if $\dot{a}$ and α are also considered constant during the time interval $\alpha/\dot{a}$, a solution can be obtained since the convolution integral in equation (9.53) is a linear convolution. Then, using the result of equations (9.40) and (9.41)

$$\Delta u_2 = E_R D(\tilde{t}) \Delta u_2^R, \tag{9.55}$$

where $\tilde{t} = ks/\dot{a}$, with $k \simeq 1/3$ (see also equation (9.41)). Substituting equation (9.55) into equation (9.54):

$$E_R D\left(\frac{k\alpha}{\dot{a}}\right) \int_0^\alpha \sigma_f \frac{\partial \Delta u_2}{\partial s} ds = 2\Psi, \tag{9.56}$$

or

$$E_R D\left(\frac{k\alpha}{\dot{a}}\right) J_v = 2\Psi. \tag{9.57}$$

Note that J_v is the energy release rate of the reference elastic continuum and applied forces into the crack.

Power-law nonlinearity assists in the analysis. An introduction was given in Section 8.2. Schapery (1984a) used this assumption in formulating an analysis based on dimensionless coordinates and mechanical state variables to study the crack-opening displacement for nonlinear elastic materials. This analysis arrived at the result that the variation of the dimensionless crack opening displacement across the failure zone is given by the function $g(\eta)$ where (as in the linear theory), $\eta = s/\alpha$, as was used before in equation (9.17). This seems very reasonable in fact as an assumption. Then it is found that

$$\Delta u_2^R = \alpha \left(\frac{\sigma_M}{\sigma_0}\right)^{\frac{1}{m}} g(\eta), \tag{9.58}$$

where σ_M is a normalizing constant to define a dimensionless failure stress distribution $f = \sigma_f/\sigma_M$, and σ_0 ensures a dimensionless potential $\Phi_0 = \Phi/\sigma_0$. Note that

σ_M is not necessarily the maximum stress in the failure zone, as in linear theory. The quantity $m + 1$ defines the power-law nonlinearity of the effective strain energy density Φ.

We want to take advantage of the relationship $J_f = J_v$; first we have from equation (9.50)

$$J_f = \int_0^\alpha f(\eta)\,\sigma_M \frac{\partial \Delta u_2^R}{\partial s}\,ds; \tag{9.59}$$

then from equation (9.58)

$$\frac{\partial \Delta u_2^R}{\partial s} = \alpha \left(\frac{\sigma_M}{\sigma_0}\right)^{\frac{1}{m}} \frac{\partial g\left(\eta = \frac{s}{\alpha}\right)}{\partial s}\,ds = \alpha \left(\frac{\sigma_M}{\sigma_0}\right)^{\frac{1}{m}} \frac{\partial g(\eta)}{\partial \eta}\,d\eta. \tag{9.60}$$

Inserting this into equation (9.59):

$$J_f = \alpha \left(\frac{\sigma_M}{\sigma_0}\right)^{\frac{1}{m}} \sigma_M \int_0^\alpha f(\eta) \frac{\partial g}{\partial \eta}\,d\eta, \tag{9.61}$$

and using $J_f = J_v$ we can solve for α :

$$\alpha = \left(\frac{\sigma_0}{\sigma_M}\right)^{\frac{1}{m}} \frac{J_v}{\sigma_M I_f}, \tag{9.62}$$

where

$$I_f = \int_0^1 f(\eta) \frac{dg}{d\eta}\,d\eta. \tag{9.63}$$

We see a similar form to that in the linear solution for α in equation (9.18). From equations (9.58) and (9.62) we have

$$\Delta u_2^R = \left(\frac{J_v}{\sigma_M I_f}\right) g(\eta). \tag{9.64}$$

The viscoelastic solutions follow from the correspondence principles; for example, from equation (9.64),

$$\Delta u_2 = E_R \int_0^t D(t - \tau) \frac{\partial \Delta u_2^R}{\partial \tau}\,d\tau. \tag{9.65}$$

This is similar to equation (9.53), and provides the basis for further analysis.

Now we consider power-law viscoelasticity $D(t) = D_0 t^n$. We also assume σ_m and Ψ constant. Then equations (9.57) and (9.62) can be used to show that the crack speed is given by

$$\dot{a} \propto J_v^p, \tag{9.66}$$

where $p = 1 + (1/n)$ is a constant. Equation (9.57) can be used to solve for crack speed in terms of process zone length, which in turn is related to J_v via equation (9.62). This is an important result for use in engineering studies of ice. The relationship was used in Jordaan and Xiao (1992) to study the time-to-failure for a growing crack.

9.6   Experimental Comparisons: Beam Bending and Compact Tension Experiments

Figure 2.9 shows an almost three-fold decrease in the apparent fracture toughness with loading rate. The results were well fitted with a power-law decreasing function by Kavanagh (2018), who reviewed past work on time-dependent measurements of ice fracture and conducted further tests and analyses of time-dependency of ice fracture. His analysis is based largely on the linear viscoelastic theory (Section 9.4); see also Kavanagh and Jordaan (2022). Two test series were conducted on beams using granular polycrystalline ice, and methods for specimen preparation generally following those outlined in Appendix B. A four-point loading was used as illustrated in Figure 9.7. The beam width was 60 mm, thickness 40 mm and length S_2 was 285.75 mm. The distance from the supports to the load points was 92.875 mm so that S_1 was 100 mm. The crack length was 10 mm. Each ice sample was notched with a fine-tooth saw to a depth of 10 mm, and a razor blade was used to give the crack a sharp edge. Prior to each test, the razor blade was run through the crack to prevent the cracks from healing and becoming dull. Further details can be found in Kavanagh (2018). He also analysed the test results of Liu and Miller (1979) in which the compact tension

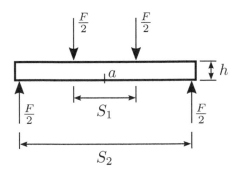

Figure 9.7 Test set-up for 4-point bending fracture tests.

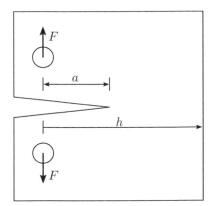

Figure 9.8 Schematic of test set-up for compact tension experiment used by Liu and Miller (1979). With permission from Cambridge University Press.

experiment set-up was used (shown schematically in Figure 9.8). The specimens were 122 mm by 127 mm (the latter in the direction of the crack), with thickness 25.4 mm. The distance h (Figure 9.8) was 104 mm, and the crack length a was 46.5 mm.

The objective was to test the theory using reasonably determined input values. The linear elastic theory developed by Barenblatt was based on consideration of an idealized crack in an infinite body under symmetrical loading; Schapery based his solution on the equations developed by Williams (1957). To make this solution developed by Schapery in equation (9.45) to be applicable to these situations, several factors needed to be studied. First, the geometric situation is different in the new four-point bending configuration and in the compact tension so that the fracture toughness (via the geometric shape factor) was updated to reflect the 4-point loading conditions; see for example the references Strecker et al. (2005) and Fett (2008); correct usage is noted in Kavanagh and Jordaan (2022). In a time-stepping analysis, the model was implemented with an explicit Euler forward method. Both crack initiation and growth were considered with growth constituting the most important aspect. Factors such as geometric shape function, stress intensity, effective compliance, failure zone size, fracture toughness and finally crack length were updated in the algorithm until failure of the specimen is obtained.

Schapery's method requires approximation of the creep compliance of equation (9.43). The effective compliance, discussed in equations (9.30) to (9.33), defines the correspondence between the elastic and viscoelastic solutions. Kavanagh analysed a number of data sets, including LeClair et al. (1999) and Brill and Camp (1961). The results showed that creep curves for ice could be well represented by the effective compliance and that the linear approximation is satisfactory up to times possibly as high as 100 s. The tests of Brill and Camp were most promising since they used polycrystalline ice with a grain size similar to Kavanagh's tests; their tests were conducted at −5°C, somewhat warmer than Kavanagh's tests at −10°C. The results were used to determine the constants C_0 and C_1 in equation (9.43), which could then be used to solve the viscoelastic equations using the effective compliance of equations (9.33)

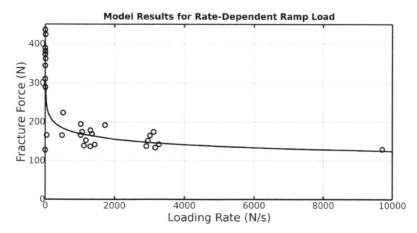

Figure 9.9 Comparison of theory and experiment for Kavanagh's test results. With permission.

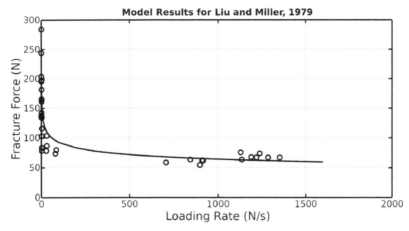

Figure 9.10 Comparison of theory and experiment for test results of Liu and Miller (1979). Prepared with permission from Cambridge University Press.

and (9.34). In addition to the creep-related parameters, other parameters such as σ_M and the specific fracture energy Ψ are needed. Comparisons of the theory with experimental results using the parameters from the Brill and Camp tests yielded very good results. We consider now the tests with a range of loading rates. Using the parameters $C_0 = 1.20 \times 10^{-10}$ Pa^{-1}, $C_1 = 1.44 \times 10^{-10}$ Pa^{-1}s$^{-0.313}$, $n = 0.313$, $\sigma_m = 318$ MPa, $\Psi = 1.61$ Jm^{-2}, the results in Figure 9.9 were obtained. Similar analyses were carried out for the tests of Liu and Miller (1979), with modified input parameters. The results are shown in Figure 9.10.

In conclusion, an approach to time-dependent cracking of ice has been developed. The approach is promising with realistic process zone sizes. Certainly more needs to be done: simplifications of the approach are needed, yet the solutions provided are in the realm of appropriate viscoelastic theory to a problem that is essentially time-dependent. We believe that a rich ground for further research has been presented.

10 Concluding Remarks

This work relates how ice mechanics, as applied to offshore engineering, has developed in the last 50 years to the point where a good understanding of ice pressures, both local and global, on offshore structures has been obtained. The focus of the work has been on interaction with thick ice features and icebergs. The progress results from Canada's past in having to contend with its vast northern oceans including transportation of goods as well as construction of installations in ice-covered waters, and in particular exploration of its northern regions for resources during the 1970s and 1980s. This led to a series of imaginative experiments such as the use of instrumented icebreakers in ramming of hard ice features and iceberg bergy bits, the Hans Island experiments, and the experiments involving field medium scale indentation of ice. Significant efforts over the years have been devoted to ice engineering, and a strong flourishing of knowledge has resulted, with new industrial developments, especially in the marginal areas offshore Newfoundland, where several fixed and floating structures operate in waters frequented by icebergs.

Understanding of the mechanics of ice failure has progressed, but often not to the point of being of direct numerical use in design. Armed with background knowledge, there is enough empirical data which, combined with probabilistic analysis, affords a reasonable approach to design. Approaches made on this basis have been summarized in Part I of this work. A most important factor in compressive ice failure is the occurrence of high-pressure zones (*hpz*s). At very low interaction speeds, the distribution of ice stresses against the structure is relatively uniform, but at higher rates crushing occurs with localization of pressure into *hpz*s. These zones may constitute only a small proportion—of the order of 10%—of the global interaction area. Although fracture is involved in this process, a full explanation based on fracture mechanics seems a long way off. The behaviour of ice within the *hpz*s involves further localization into a layer of microstructurally modified material, whose mechanical response is quite different, by many orders of magnitude, from that of the parent undamaged ice. It is noted that the distribution of pressure across the *hpz*s is rather irregular, reflecting local fracture events and irregularities in ice structure, which need further analysis.

Above all, the importance of time in the processes described is paramount. Consequently the appoach to mechanics in Part II of this work has been based on viscoelastic methods coupled with damage mechanics, the latter to reflect the microstructural changes to the ice and consequent changes to the viscoelastic response. An analysis of layer development has been presented based on these principles. Strong emphasis has been placed on the work of R. A. Schapery, as his approach is highly relevant to the problems of ice mechanics. The creep of ice under uniaxial stresses is well known and documented, but the states of stress in uniaxial loading at low stresses are a far cry from those in the layer and its vicinity. As a result, the research programme required an extensive experimental exploration of ice response under triaxial stress states focussed on the problem at hand. Behaviour of the ice was found to be highly nonlinear and strongly dependent on the hydrostatic pressure. Finite element simulations in ABAQUS using the user-material option have shown promising results. The simulations included modelling based on the triaxial experiments that shown a "stiffening" effect as pressures increase, yet with significant softening associated with pressure melting effects at even higher pressures. On the question of fracture, basic measurements have been made in past research to determine fracture toughness. It is appropriate now for research to proceed to analysis of field situations. Application to real situations is needed, especially to explain the formation of *hpz*s. The question of time-dependency of fracture has been emphasized, and initial analyses based on linear viscoelastic analysis are promising. Extension to the nonlinear case has also been introduced. There is much scope for further work: the extrusion that follows layer development is one prominent example.

In the discussion of recrystallization in Section 2.6, the similarity, on vastly different length and timescales, between models for a major fault zone (e.g. Sibson, 1977) and the processes described in the book, on a much smaller scale, have been pointed out. One aspect that our tests have brought out clearly is the importance of pressure. This builds up considerably within the high-pressure zones with consequent recrystallization of the ice, and might be analogous to mylonite formation.

Appendix A Précis of the Work of R. A. Schapery

A.1   R. A. Schapery

Richard Schapery was born in Duluth, Minnesota. His paternal grandparents were immigrants from western Russia. Originally the family name was Shapiro, but his great grandfather wanted to distinguish his name from his wife's family name, which was also Shapiro. His paternal grandfather immigrated from western Russia in 1909, while his paternal grandmother arrived in 1897, also from western Russia. His mother was born in Ukraine and came to the US as an infant in 1912 with the family name Slovutsky. Both families settled in Duluth, MN, where Richard was born. His father worked for the US Corps of Engineers for most of his life, and was assigned to the St. Lawrence Seaway project during its construction.

Richard Schapery received early education at Wayne State University, and a Ph.D. from the California Institute of Technology in 1962. He served at Texas A & M University, College Station, Texas, USA, for some years and he was then the Cockrell Family Regents Chair in Engineering and Professor of Aerospace Engineering and Engineering Mechanics at University of Texas at Austin. He retired from this department in August 2002. He specializes in mechanics of materials, primarily in the development of mathematical modelling of material behaviour and of viscoelastic analysis. His contribution to these areas has been immense. He combines profound knowledge of fundamental theory and thermodynamics with a keen sense of the requirements for practical analysis and design, all with great attention to detail.

A.2   Introductory Comments

In our analysis of viscoelasticity we use the ideas of adiabatic and isothermal conditions. Under isothermal conditions, there will be a small increase in temperature when a solid is compressed, and conversely a decrease if subjected to tension. Sliding displacements associated with movement of dislocations in creep, for example, lead to the generation of heat and changes of temperature under adiabatic conditions. We shall assume that this heat will be removed or added by immersing the specimen in a very large reservoir (heat-bath) at constant temperature, thus leading to isothermal conditions.

The effects of this assumption upon elastic properties is small (Landau and Lifshitz, 1970). The following summarizes the relation between the elastic constants K, G, and E, measured under isothermal circumstances, with K_{ad}, G_{ad} and E_{ad} being the corresponding adiabatic ones.

$$\frac{1}{K_{ad}} = \frac{1}{K} - \frac{T\beta^2}{C_p}; G_{ad} = G; E_{ad} = \frac{E}{1 - \frac{ET\beta^2}{9C_p}} \simeq E\left(1 + \frac{ET\beta^2}{9C_p}\right), \qquad (A.1)$$

where β is the coefficient of volume expansion. This is approximately three times the coefficient of linear expansion. It is to be noted that in these equations C_p is the specific heat per unit volume. The difference between isothermal and adiabatic values is relatively small. The reader may wish to check the following result. Given the following values for ice at $-10°C$: $C_p = 203$ J g^{-1} °C^{-1}, $\beta = 1.55 \times 10^{-6}$°C^{-1}, find K_{ad}/K. First, consider 1 m^3 of ice: $C_p = 2.214$ J m^{-3} °C^{-1} (conversion per mass to per volume) and we find by using equation (A.1) that $K_{ad}/K = 1.0223$, or the adiabatic bulk modulus is 2.23% greater than the isothermal value.

A.3 Biot–Schapery Theory

The approach taken in the present work is based on that of Biot (1954, 1958) and Schapery (1964, 1968). While being soundly based on the thermodynamics of irreversible processes (TIP), it does at the same time offer well-based solutions to practical problems. We shall not review the extensive literature on the foundations of viscoelastic theory, but rather state that the work of the two authors, while being soundly based, leads to formulations that are of immediate use in practical problems. Schapery (1991a) notes that the derivation of Biot (1954) leads to "the most general form of time-dependence physically possible".

Our focus in the present context is on viscoelastic relationships to be used mainly in compression, together with the effect of concurrent damage. The sign convention will be positive for compression and contraction, but relevant parts of the methodology are quite general and can be used with tension and extension being positive, as was done in Chapter 9, when considering fracture. The use of generalized forces Q_i and associated generalized displacements q_i gives useful context to the use of state variables. One can consider these as representing pressure p and volume v, or stress σ_{ij} and strain ϵ_{ij}, respectively. But there are several other additional functions of the $\{Q_i, q_i\}$ set, including, for example, hidden coordinates, internal tractions and damage. We shall include hidden variables in the present analysis and deal with damage in Section A.5. The internal variables or hidden coordinates account for effects such as microscopic structural rearrangements, crystal slip and grain boundary sliding. The observed coordinates are denoted by subscripts $i = 1, 2, \ldots, k$, and the hidden set by $i = k + 1, \ldots, n$. The hidden coordinates are characterized by the fact that their associated generalized forces are zero. They appear in the analysis eventually as components of the viscoelastic compliance, for instance as a set of Kelvin (delayed elastic) units, where each internal unit has no external force applied.

We now consider the work done by keeping the Q_i's constant and varying the displacements by an increment δq_i, such that the product $Q_i \delta q_i$ (no summation implied) is an increment of work. For instance, for a material under hydrostatic pressure p, the increment of work is $p \delta v$; for a generally stressed solid, the increment is $\sigma_{ij} \delta \epsilon_{ij}$ (again, no summation). Note that forces are kept constant and displacements varied. Our system Σ of interest is the ice specimen which we place in a large heat reservoir, large enough to maintain isothermal conditions. We divide the system into two subsystems: Σ_1, the specimen, and Σ_2, the reservoir.

Now applying the first law of thermodynamics to Σ_1, the equation ($\Delta u = \Delta h + \Delta w$) becomes

$$\dot{u} = Q_i \dot{q}_i - \dot{h}, \tag{A.2}$$

where summation is implied, and $\dot{h}$ is the rate of heat leaving system Σ_1 (the specimen of ice) and absorbed by the heat bath (system Σ_2). This equation states that the rate of work applied to the specimen ($Q_i \dot{q}_i$) either becomes stored as internal energy ($\dot{u}$) or is dissipated ($\dot{h}$). The entropy production rate is composed of two parts:

$$\dot{S} = \dot{S}_1 + \dot{S}_2, \tag{A.3}$$

where the two rates $\dot{S}_1$ and $\dot{S}_2$ correspond to subsystems Σ_1 and Σ_2. But we know that the heat $\dot{h}$ is transferred from Σ_1 to Σ_2, so that

$$\dot{S}_2 = \frac{\dot{h}}{T}, \tag{A.4}$$

from equation (2.2). Then from equations (A.2), (A.3) and (A.4):

$$T\dot{S} = T\dot{S}_1 + \dot{h}, \text{ and} \tag{A.5}$$

$$T\dot{S} = T\dot{S}_1 + Q_i \dot{q}_i - \dot{u}. \tag{A.6}$$

We now introduce the Helmholtz free energy F:

$$F = u - TS, \tag{A.7}$$

so that equation (A.6) becomes

$$T\dot{S} = Q_i \dot{q}_i - \dot{F}. \tag{A.8}$$

We can interpret this equation as the entropy being produced, for example, through the movement of dashpots when the work done ($Q_i \dot{q}_i$) exceeds the work associated with the "reversible force", defined as

$$Q_i^R = \frac{\partial F}{\partial q_i}, \tag{A.9}$$

absorbing the deformation $\dot{q}_i$. Thus

$$T\dot{S} = \left(Q_i - Q_i^R \right) \dot{q}_i = X_i \dot{q}_i. \tag{A.10}$$

This is illustrated conceptually in Figure A.1.

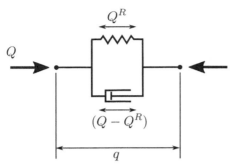

Figure A.1 Illustration of dissipation and storage as described in equation (A.10).

Schapery (1964, 1968) expanded the free energy into a Taylor series at the reference state, but neglecting all terms higher than second order:

$$F = \left(\frac{\partial F}{\partial T}\right)_r \theta + \left(\frac{\partial F}{\partial q_j}\right)_r q_j + \frac{1}{2}\left(\frac{\partial^2 F}{\partial q_i \partial q_j}\right)_r q_i q_j + \left(\frac{\partial^2 F}{\partial q_j \partial T}\right)_r \theta q_j + \frac{1}{2}\left(\frac{\partial^2 F}{\partial T^2}\right)_r \theta^2, \quad (A.11)$$

in which $\theta = T - T_r$; q_i and F are arbitrarily taken as zero at the reference state (r). In the reference state, all the Q_i are zero, and the temperature is T_r. By definition of the reference state

$$\left(\frac{\partial F}{\partial q_i}\right)_r = 0. \quad (A.12)$$

We now have

$$F = -S_r\theta + \frac{1}{2}a_{ij}q_i q_j - \beta_i \theta q_i - (C_q/2T_r)\theta^2, \quad (A.13)$$

where

$$a_{ij} = a_{ji} = \left(\frac{\partial^2 F}{\partial q_i \partial q_j}\right)_r, \beta_i = -\left(\frac{\partial^2 F}{\partial q_i \partial T}\right)_r. \quad (A.14)$$

The matrix a_{ij} is positive semi-definite; S_r and C_q are the entropy and the specific heat at the reference state, respectively. Now, following Schapery, we write

$$X_i = Q_i - Q_i^R = Q_i - \frac{\partial F}{\partial q_i} = b'_{ij}\frac{dq_j}{dt} = a_D b_{ij}\frac{dq_j}{dt}, \quad (A.15)$$

in which b'_{ij} is a positive semi-definite matrix, and a_D is an entropy production coefficient. For instance, in the case of stress-dependent nonlinear creep (Glen's law, Sections 2.4 and 6.8), we have

$$a_D = \mu(\sigma) \propto \frac{1}{\sigma^{n-1}}. \quad (A.16)$$

The coefficient a_D could also be a function of temperature, for example.

The matrices b'_{ij} and b_{ij} are symmetrical, heuristically, our preference, or by Onsager's principle. Then

$$\frac{\partial F}{\partial q_i} + a_D b_{ij}\frac{dq_j}{dt} = Q_i. \quad (A.17)$$

Substituting equation (A.13) into equation (A.17), we have

$$a_{ij}q_j + b_{ij}\frac{dq_j}{d\psi} = Q_i + \beta_i\theta, \tag{A.18}$$

where

$$d\psi = \frac{dt}{a_D} \tag{A.19}$$

defines a "reduced time" or "pseudotime". See also Section 6.8. Note that $\{t,\tau\}$ in real time becomes $\{\psi,\psi'\}$ in pseudotime:

$$\psi = \int_0^t \frac{dt'}{a_D}; \psi' = \int_0^\tau \frac{dt'}{a_D}. \tag{A.20}$$

We now consider isothermal conditions and outline the method of solution (Biot and Schapery; see also Fung, 1965) to obtain the q_i as a function of the Q's. The method consists of simultaneous diagonalization of the matrices a_{ij} and b_{ij}, thus uncoupling the equations of motion. We may write equation (A.18) in operational form:

$$\sum_j A_{ij}q_j = Q_i, \tag{A.21}$$

where

$$A_{ij} = a_{ij} + pb_{ij}, \tag{A.22}$$

and

$$p \equiv \frac{d}{d\psi}. \tag{A.23}$$

Equation (A.21) can first be thought of as a set of homogeneous linear equations, with p considered for present purposes as a parameter $(-\lambda)$ (Biot, 1954; see also Fung, 1965). We consider first the homogenous case

$$\sum_j A_{ij}q_j = 0. \tag{A.24}$$

This leads to an eigenvalue problem in which the characteristic values are obtained by setting the determinant

$$\det\left(A_{ij}\right) = \det\left(a_{ij} - \lambda b_{ij}\right) = 0. \tag{A.25}$$

The solution of equation (A.24) is obtained (see Fung (1965) for details) as

$$q_i = \sum_k \varphi_j^{(k)} \exp\left(-\lambda_k\psi\right), \tag{A.26}$$

where the λ_k are eigenvalues with associated eigenvectors $\varphi_j^{(k)}$ which are linearly independent and orthonormal. The solutions for each λ_k,

$$q_j^{(k)} = \varphi_j^{(k)} \exp(-\lambda_k\psi) \tag{A.27}$$

is termed a "relaxation mode".

Proceeding now to the nonhomogeneous case, the state variables $\{q_i\}$ may be transformed linearly to the normal coordinates, say $\{\xi_i\}$, using the equations

$$q_i = \sum_k \varphi_i^{(k)}\xi_k, \tag{A.28}$$

where $\left\{\varphi_i^{(k)}\right\}$ is the modal column of the kth relaxation mode. We can then write that

$$\dot{\xi}_k + \lambda_k \xi_k = \Xi_k, \tag{A.29}$$

corresponding to equation (A.21) but decoupled, and Ξ_k is the corresponding generalized force. Taking the Laplace transform:

$$(s + \lambda_k)\overline{\xi}_k = \overline{\Xi}_k, \tag{A.30}$$

and

$$\overline{\xi}_k = \frac{\overline{\Xi}_k}{s + \lambda_k}. \tag{A.31}$$

We may now substitute these values into the Laplace-transformed equation (A.28):

$$\overline{q}_i(s) = \sum_j \overline{Q}_j(s)\left[\sum_k \frac{C_{ij}^{(k)}}{s + \lambda_k} + C_{ij}\right], \tag{A.32}$$

where

$$C_{ij}^{(k)} = \varphi_i^{(k)}\varphi_j^{(k)}, \tag{A.33}$$

and C_{ij} corresponds to infinite roots (single or multiple). The solution in time (or here, pseudotime) is found in the inverse Laplace transform. This is as follows:

$$q_i(\psi) = \sum_j \left\{\sum_k C_{ij}^k \int_0^\psi \exp\left[-\lambda_k(\psi - \psi')\right] Q_j(\psi')\,d\psi' + C_{ij}Q_j(\psi)\right\}; \tag{A.34}$$

recall that pseudotime is represented by

$$\psi = \int_0^t \frac{dt'}{a_D}; \psi' = \int_0^\tau \frac{dt'}{a_D}. \tag{A.35}$$

These equations are beginning to look like the equations of linear viscoelasticity (Chapter 6), and we wish to solve for the applied forces in terms of the observed coordinates. This is achieved by inverting the matrix equations (A.34) in terms of the observed coordinates and eliminating the hidden ones, yet maintaining their influence on the model. In the following matrix representation of equation (A.21), the forces Q_i are applied to the k coordinates $q_1, \ldots, q_k$ while the remaining $n - k$ coordinates comprise the internal system. The latter is represented by the symmetric square submatrix $[M]$ as shown. We have assumed that $[M]$ has been diagonalized using $\det[M] = 0$, and the hidden coordinates are then represented by the normal coordinates ξ_j, $j = k+1, \ldots, n$, and primes attached to the appropriate entries in the matrix to show that the values have been adjusted. The procedure is to solve for the hidden coordinates in terms of the observed ones and to substitute the values back into the first k equations. Details of the procedures can be found in the papers Biot (1954, 1958) and Schapery (1964, 1968).

$$
\begin{bmatrix}
A_{11} & \cdots & A_{1k} & | & A'_{1,k+1} & \cdots & A'_{1n} \\
\vdots & & \vdots & | & \vdots & & \vdots \\
A_{k1} & \cdots & A_{kk} & | & A'_{k,k+1} & \cdots & A'_{kn} \\
-- & -- & -- & -- & -- & -- & -- \\
A'_{k+1,1} & \cdots & A'_{k+1,k} & | & & & \\
\vdots & & \vdots & | & & M & \\
A'_{n1} & \cdots & A'_{nk} & | & & &
\end{bmatrix}
\begin{bmatrix}
q \\ \vdots \\ q_k \\ -- \\ \xi_{k+1} \\ \vdots \\ \xi_n
\end{bmatrix}
=
\begin{bmatrix}
Q_1 \\ \vdots \\ Q_k \\ -- \\ 0 \\ 0 \\ 0
\end{bmatrix}. \tag{A.36}
$$

When this has been done, we obtain, again in operational form, here using the Laplace transform, as in Fung (1965):

$$
\overline{Q}_i = \sum_j T_{ij} \overline{q}_j, \tag{A.37}
$$

where

$$
T_{ij} = \sum_m \frac{D_{ij}^{(m)}}{s + \lambda_m} + D_{ij} + s D'_{ij}, \tag{A.38}
$$

$$
D_{ij}^{(m)} = \Phi_i^{(m)} \Phi_j^{(m)} \tag{A.39}
$$

and the Φ's, D_{ij} and D'_{ij} relate to the transformation of coordinates. The final result is

$$
Q_i(\psi) = \sum_j \left\{ D_{ij} q_j(t) + D'_{ij} \dot{q}_j + \sum_m D_{ij}^{(m)} \exp\left[-\lambda_m (\psi - \psi') q_j(\psi) d\psi \right] \right\}. \tag{A.40}
$$

(Matrices are positive semi-definite.) This is the equation of the Maxwell model comprised of springs and dashpots, as shown in Figure 6.9. The first proof was given by Biot for linear systems; Schapery added the entropy production coefficient which permitted nonlinearities, and also the effect of temperature and further nonlinearity as we now describe.

In Schapery (1966) it was shown that the analysis included consideration of nonlinear springs and dashpots rather than the linear ones previously considered. He also showed (Schapery, 1968) that the generalized forces and displacements were consistent with analysis for small and large strains. For generalized coordinates q_{ab} (these are denoted as tensor functions of the Lagrangian strain tensor), the results could be written in the familiar deviatoric and dilatational forms such as the following for initially isotropic media:

$$
Q'_{ab} = 2 \int_0^t G(\psi - \psi') \frac{dq_{ab}}{d\tau} d\tau, \tag{A.41}
$$

and

$$
Q_{aa} = 3 \int_0^t K(\psi - \psi') \frac{dq_{aa}}{d\tau} d\tau. \tag{A.42}
$$

The inverses are similarly obtained. In his 1969 paper, he used Gibbs free energy in addition to the Helmholtz function and derived the generalized Kelvin (Voigt) model, as shown in Figure 6.9, but with nonlinear springs and dashpots.

The fundamental analysis outlined above provides the theoretical basis for viscoelastic theory, and is a stupendous achievement.

A.4   Constitutive Theory Using Modified Superposition Principle (MSP)

Schapery (1981, 1984a) recognized the importance of analysing and tracking energy in developing fracture and damage theory, and, at the same time, the importance of obtaining solutions that recognized time-dependence for nonlinear viscoelastic materials. He proposed a constitutive model, which had the great merit of having a related correspondence principle that permitted the derivation of nonlinear viscoelastic solutions from the associated (nonlinear) elastic ones. The theory is consistent with the analysis based on irreversible thermodynamics (TIP) given in Section A.3. We give an outline here and note that our analysis has been applied to damage associated with microcracking and with recrystallization, a related process, as described in Section A.5. An extension to large deformations and elaboration of this theory is given in Schapery (1984b); this is not dealt with here.

The MSP theory was introduced in Section 6.11 for the one-dimensional case, and used in Section 9.5, where the use of the J integral, based as it is on nonlinear elasticity, was shown. The viscoelastic strain tensor ϵ_{ij} is given by

$$\epsilon_{ij}(t) = E_R \int_0^t D(t - \tau) \frac{\partial \varrho_{ij}}{\partial \tau} d\tau, \tag{A.43}$$

where E_R is a reference elastic modulus with dimensions of stress (MPa), $D(t - \tau)$ is a linear creep compliance with dimensions of strain per unit stress. But it is to be noted that the relationship (A.43) deals with nonlinearities through the tensor ϱ_{ij}, the components of which were termed "pseudostrain" by Schapery, with the dimensions of strain. We shall term this the "effective elastic strain", as it becomes the elastic strain in the correspondence principle. This is a function of stress σ_{ij}, at position in the continuum denoted x, and time:

$$\varrho_{ij} = \varrho_{ij}(\sigma_{kl}, x, t), \tag{A.44}$$

with k and l taking values 1, 2 and 3. We emphasize again that the method is designed to relate (nonlinear) elastic J integral methods to (nonlinear) viscoelastic ones. In the correspondence principle, the tensors $\{\sigma_{ij}, \varrho_{ij}\}$ correspond to the elastic solution, while the pair $\{\sigma_{ij}(t), \epsilon_{ij}(t)\}$ corresponds to the viscoelastic solution at time t. The same stress σ_{ij} applies in both cases but the viscoelastic strain is given in equation (A.43).

For an entirely elastic relationship using equation (A.43), we put

$$D(t - \tau) = \frac{1}{E_R}; \tag{A.45}$$

note that the resulting relationship $\epsilon_{ij} = \varrho_{ij}$ does not have to be linear. A linear viscoelastic relationship with a constant Poisson's ratio v for an isotropic, homogeneous material is obtained if the effective elastic strain is given by

$$\varrho_{ij} = \frac{1}{E_R} \left[(1 + \nu)\,\sigma_{ij} - \nu\sigma_{kk}\delta_{ij} \right], \tag{A.46}$$

where δ_{ij} is the Kronecker delta. The standard uniaxial relationship for strain due to stress σ_{11} (other stress components being zero) is given by

$$\epsilon_{11} = \int_0^t D\,(t - \tau)\,\frac{d\sigma_{11}}{d\tau}\,d\tau, \tag{A.47}$$

where $\varrho_{ij} = \sigma_{ij}$. For a unit step function of value σ_0 applied at $t = 0$, $\epsilon_{11} = \sigma_0 D\,(t)$.

The function $D\,(t)$ is the usual creep compliance in linear theory, and here the usage is carried over into this new theory which includes nonlinearity via ϱ_{ij}. Ageing is included in Schapery's formulation; it is not necessary for our discussion of ice and is not included here. We shall be including a treatment of damage in the following, which is the focus of our analysis. It should also be noted that equation (A.43) can be inverted to the following form

$$\varrho_{ij} = \frac{1}{E_R} \int_0^t E\,(t - \tau)\,\frac{\partial\epsilon_{ij}}{\partial\tau}\,d\tau, \tag{A.48}$$

where $E\,(t - \tau)$ is the relaxation modulus. By taking the Laplace transforms of equations (A.43) and (A.48), one can find that

$$\overline{D}\,(s)\,\overline{E}\,(s) = \frac{1}{s^2}; \tag{A.49}$$

compare equation (6.39). This may be inverted to give the following pair of equations:

$$\int_0^t D\,(t - \tau)\,\frac{dE\,(\tau)}{d\tau}\,d\tau = H\,(t)\,; \tag{A.50}$$

$$\int_0^t E\,(t - \tau)\,\frac{dD\,(\tau)}{d\tau}\,d\tau = H\,(t)\,, \tag{A.51}$$

where $H\,(t)$ is the unit step function.

Another special case of the theory using the effective elastic strain ϱ_{ij} is that of viscous flow. Here one puts

$$D\,(t - \tau) = \frac{1}{E_R t_v}\,(t - \tau), \tag{A.52}$$

where t_v is a constant. Then

$$\epsilon_{ij}\,(t) = \frac{1}{t_v} \int_0^t (t - \tau)\,\frac{d\varrho_{ij}\,(\tau)}{d\tau}\,d\tau; \tag{A.53}$$

integrating by parts, we have

$$\epsilon_{ij}\,(t) = \frac{1}{t_v}\{\left[(t - \tau)\,\varrho_{ij}\,(\tau)\right]_0^t + \int_0^t \varrho_{ij}\,(\tau)\,d\tau\}. \tag{A.54}$$

The first term on the right hand side is zero and we then can differentiate to obtain

$$\frac{\partial\epsilon_{ij}\,(t)}{\partial t} = \frac{1}{t_v}\varrho_{ij}\,(t)\,; \tag{A.55}$$

in other words, the current stress, via ϱ_{ij} is a function of the current strain rate. As an example, consider the relationship (Glen's law)

$$\frac{\dot{\epsilon}}{\epsilon_0} = \left(\frac{\sigma}{\sigma_0}\right)^n;$$ (A.56)

here

$$t_v = \frac{1}{\epsilon_0}; \varrho = \left(\frac{\sigma}{\sigma_0}\right)^n.$$ (A.57)

See also Section 6.11.

The relationship in equation (A.43) is a good basis for studies of ice fracture and damage. We shall briefly introduce the correspondence principle upon which the analysis of deformation of a damaged continuum and fracture analysis are based. The classical correspondence principle between linear elastic media and their linear viscoelastic counterparts has been discussed in Section 6.9. Here we introduce the correspondence principle derived by Schapery for fracture and damage analysis, accounting for stationary or growing cracks. In the correspondence principle, the reference elastic solution is denoted by the superscript R, and is obtained by putting $D^{-1}(t - \tau) = E(t - \tau) = E_R$. The instantaneous geometry is the same for the reference elastic and viscoelastic solutions, and this geometry includes crack surfaces. As noted earlier, the tensors $\{\sigma_{ij} = \sigma_{ij}^R, \varrho_{ij} = \epsilon_{ij}^R\}$ correspond to the elastic solution, while the pair $\{\sigma_{ij}(t) \equiv \sigma_{ij}^R, \epsilon_{ij}(t)\}$ corresponds to the viscoelastic solution at time t. The viscoelastic strains are given by equation (A.43) while the viscoelastic displacements are given by

$$u_i = E_R \int_0^t D(t - \tau) \frac{\partial u_i^R}{\partial \tau} d\tau.$$ (A.58)

The proof rests largely on a consideration of the equilibrium and strain-displacement equations, as well as the boundary conditions, for the reference elastic solution. We write the first two of these here with the superscript R in which we have omitted consideration of body forces (which are treated in the original work, but not needed here):

$$\sigma_{ij,j}^R(x) = 0.$$ (A.59)

$$\epsilon_{ij}^R = \frac{1}{2}\left[u_{i,j}^R + u_{j,i}^R\right].$$ (A.60)

The boundary conditions are surface tractions $T_i = \sigma_{ij}n_j$ on all surfaces. It is seen that all quantities with the superscript R satisfy the equilibrium and compatibility conditions, and so do the viscoelastic stresses since these are equal to the reference values. The viscoelastic strains are obtained using the convolution integral (6.78), which is a linear operator independent of position, so that the derived strains must satisfy the compatibility conditions as well. Schapery (1981) discusses an additional correspondence principle for displacement conditions on the boundary and gives detail on the modelling of growing cracks.

A.5   Consideration of Damage

The basis of the theory with regard to damage rests on the J integral formulation. As has become abundantly clear in the analysis of ice behaviour in real compressive interactions, two kinds of damage occur: microcracking accompanied by recrystallization at low hydrostatic pressures, decreasing in importance as pressure increases, and superseded by recrystallization as the dominant mechanism of microstructural change at high hydrostatic pressures. The present section outlines the development of damage theory based on the Schapery (1981) J integral formulation. It is believed that this analysis can be applied to both of the mechanisms discussed; it was developed primarily with microcracking in mind. Yet we note that recrystallization also is related to factors such as dislocation density; no doubt other irregularities play a role: pores, cracks and grain boundaries where the crystallographic direction of the lattice abruptly changes (as occurs when two crystals begin growing separately and then meet). We note also that Eshelby (1971) terms the J integral as the "effective force on crack tip" and states that "energy release rate is closely related to the force on a defect (dislocation, impurity atom, lattice vacancy and so forth)" in the theory of lattice defects.

In Section 9.5, we introduced Schapery's J integral analysis with regard to crack formation under tensile states. We now consider damage under compressive states of stress. To use the J integral method, we must characterize the reference material as a nonlinear elastic material for which a potential Φ exists such that

$$\sigma_{ij} = \frac{\partial \Phi}{\partial \varrho_{ij}}, \tag{A.61}$$

and which should be interpreted as an "effective strain energy density". In addition, a complementary effective potential Φ_c exists, such that

$$\varrho_{ij} = \frac{\partial \Phi_c}{\partial \sigma_{ij}}. \tag{A.62}$$

The usual relationship between the two potentials applies:

$$\Phi_c = -\Phi + \sigma_{ij}\varrho_{ij}. \tag{A.63}$$

The methodology needs to account for damaged material, microcracked or recrystallized zones or layers. If we consider a mass of damaged ice, in addition to the external surfaces, there will be regions of intense damage for which the potentials of the parent material do not apply. As a result, the damaged material includes such zones in addition to the parent material, and further, there will be tractions acting between these zones, contributing to surface tractions T_i. Then the surface S is defined as the bounding surface of material obeying the constitutive equations and equation (A.61). Schapery (1981) re-formulated the problem and showed that potentials of the kind just noted can be derived for the stresses on the external surface (denoted S_e) only. We shall therefore continue to use equations (A.61) to (A.63) for this revised case, and add a superscript R to denote the reference elastic solution for the problem at hand. These are relationships applying to the reference state and are here termed "effective potentials".

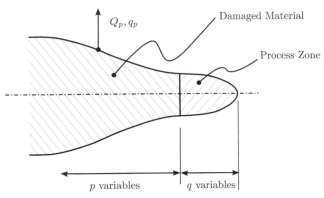

Figure A.2 Conceptual illustration of parameters relevant to damage growth. Note that p and q variables constitute the (r, s) set.

In order to take into account the microstructural changes in analysing the reference elastic problem, Schapery extended the description in Section A.3. The unit volume of material is characterized by the effective potential Φ^R and is a function of generalized displacements $q_\alpha (\alpha = 1, 2, \ldots, N)$. The N generalized forces Q_α and associated displacements q_α are as defined in Section A.3. The generalized coordinates are divided into two groups: $\alpha = 1, 2, \ldots, 6$, representing the stresses and strains applied to the external surface S_e (also denoted as σ_{ij} and ϱ_{ij}). The remaining coordinates $\alpha = 7, 8, \ldots, N$ denote quantities related to the microstructural changes, crack length, crack surface displacements as well as those related to dynamic recrystallization. The subscripts $m, n = 1, \ldots, 6$ relate to the first set while $r, s = 7, \ldots, N$ relate to the second set.

The boundary between the two is illustrated in Figure A.2. Note that the p and q variables constitute the (r, s) set. We term the zone where we analyse energy flux via the J integral the "process zone", generalizing the "failure zone" for crack propagation of Section 9.5. The following discussion will be based largely on the original 1981 paper by Schapery which focussed on microcracks but we note that Schapery (1991a) discusses generalization of the processes, including for example, the number of broken bonds and the size of a cavity. Our concern is recrystallization as well as microcracking, so that we add recrystallized volume to the processes to be considered. The effective potential models the work done on the elastic material under consideration by all stresses and displacements, including tractions on crack faces. Referring to Figure A.2, the variables that are relevant to damage growth are delineated. The $\{Q_q, q_q\}$ variables refer to the processes in the process zone, essentially modelled by the J integral, while the set $\{Q_p, q_p\}$ refers to tractions of the surface of the damaged or cracked area. The process zone is modelled using J integral theory; this is shown in Figure A.3 for the case of a microcrack. Following observations, these are seen as initiating, growing and then being arrested. Mode I stresses are shown in Figure A.3, but the other modes are included in the analysis for a general stress state.

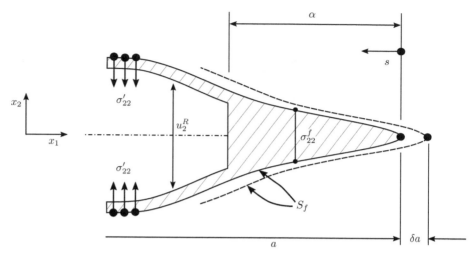

Figure A.3 Self-similar crack propagation; mode I is shown but other modes are included in analysis. Refer also to Figure 9.3. Based on Schapery (1984a) with permission from Springer Nature.

The following potential function was introduced:

$$\Phi^R = \Phi^R\left(q_m, q_r\right),\qquad\text{(A.64)}$$

with the complementary function, including tractions on the surface of the damaged area:

$$\Phi_f^R = -\Phi^R + Q_m q_m + Q_p q_p.\qquad\text{(A.65)}$$

As a result,

$$Q_m = \frac{\partial\Phi^R}{\partial q_m}, Q_q = \frac{\partial\Phi^R}{\partial q_q}; q_m = \frac{\partial\Phi_f^R}{\partial Q_m}; q_p = \frac{\partial\Phi_f^R}{\partial Q_p}.\qquad\text{(A.66)}$$

As noted, the process zone $\left\{Q_q, q_q\right\}$ pair is handled via the J integral. We summarize the approach with regard to crack propagation, noting that other processes can be dealt with in a similar way. Considering the surface of the elastic material $S_f = S - S_e$, internal tractions do work during small crack growth δa which comprises two components: first the work on the damage surfaces outside the process zone by the pair $\left\{Q_p, q_p\right\}$ and second by the flux into the process zone during the fracture or recrystallization process. We associate q_q with new crack area and find that the energy from the elastic material is

$$J^R L\delta a,\qquad\text{(A.67)}$$

where $L\delta a$ is the new crack area, and

$$J^R = J_1^R + J_2^R + J_3^R,\qquad\text{(A.68)}$$

where

$$J_1^R = \int_0^\alpha \left(\sigma_{22}^f \frac{\partial \Delta u_2^R}{\partial \xi}\right) d\xi; \ J_2^R = \int_0^\alpha \left(\sigma_{21}^f \frac{\partial \Delta u_1^R}{\partial \xi}\right) d\xi; \ J_3^R = \int_0^\alpha \left(\sigma_{23}^f \frac{\partial \Delta u_3^R}{\partial \xi}\right) d\xi,$$

(A.69)

the usual J integral terms. In Section 9.5 there is discussion of this energy transfer.

We now consider crack speed. This is analysed via the relationship between crack opening displacement and its corresponding viscoelastic counterpart; recall also that the compliance in the MSP method is linear with nonlinear potential input. Referring also to equation (9.57) and to Schapery (1984b), the crack speed for the qth segment, the crack speed is given by

$$\frac{da_q}{dt} = f_q \left(J_{1q}^R, J_{2q}^R, J_{3q}^R\right).$$

(A.70)

The analysis is assisted by using the relationships for power-law materials (see, for example, Sections 8.2, 8.4 and 9.5). We now consider proportional loading defined as follows:

$$\sigma_{ij} = \sigma \hat{\sigma}_{ij},$$

(A.71)

where $\sigma = \sigma(t)$ defines the time-dependence of stresses and $\hat{\sigma}_{ij}$ is a dimensionless constant tensor. The value $(n + 1)$ defines the power-law:

$$\Phi_f^R \left(\sigma_{ij}\right) = \Phi_f^R \left(\sigma \hat{\sigma}_{ij}\right) = \sigma^{n+1} \Phi_f^R \left(\hat{\sigma}_{ij}\right).$$

(A.72)

The viscoelastic strain is given by

$$\epsilon_{ij}^v = E_R \int_0^t D(t - \tau) \left[\sigma^n \frac{\partial \hat{\Phi}_f^R}{\partial \sigma_{ij}}\right] d\tau,$$

(A.73)

where $\hat{\Phi}$ is the potential for stress $\sigma = 1$.

Under proportional loading all stresses including those on crack faces Q_p must vary in a proportional manner. The process zone is considered to be small and the tractions in this zone do not need to be considered as proportional, as the effect on the global response is negligible. We also note again that we have been considering compressive states so that the *sign* notation has not been used.

Since

$$\Phi_f^R \propto \sigma^{n+1},$$

(A.74)

and that crack growth

$$\frac{da}{dt} = f\left(J^R, J_2^R, J_3^R\right),$$

(A.75)

also, using equation (A.66)

$$J_q^R = -\frac{\partial \Phi^R}{\partial q_q} = \frac{\partial \Phi_f^R}{\partial q_q}.$$

(A.76)

Then

$$J^R, J_i^R \propto \sigma^{n+1}.$$

(A.77)

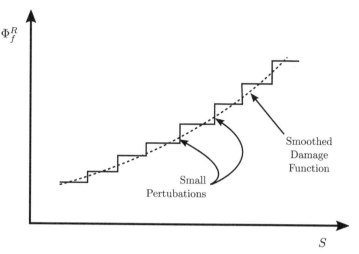

Figure A.4 Growth of damage by small steps.

Assuming a homogeneous function of degree k to model the relationship in equation (A.75):

$$\frac{da}{dt} = f_1 \left(J^R \right)^k = f_1 (f_2 \sigma^{n+1})^k, \tag{A.78}$$

with f_2 independent of stress. For example, in the case of an isolated penny-shaped crack, $f_2 = f_3 a$, where a = crack length, f_3 = constant.

Integrating (see also Jordaan and Xiao, 1992):

$$a^{1-k} - a_0^{1-k} = (1 - k) f_3^k \int_0^t f_1 \sigma^{(n+1)k} dt. \tag{A.79}$$

If $k > 1$, then $a \to \infty$ at time t_f, the "local failure time". We may write

$$\left[a_0^{k-1} (k - 1) f_3^k \right]^{-1} = \int_0^{t_f} f_1 \sigma^{(n+1)k} dt, \tag{A.80}$$

and designate this quantity as the damage parameter. In general, for the ith crack or event under consideration:

$$S_i = \int_{01}^{t_i} f_1 \sigma^{(n+1)k} dt. \tag{A.81}$$

The quantity S is the damage parameter; generally

$$S = \int_0^t f_1 \sigma^q d\tau, \tag{A.82}$$

where

$$q = (n + 1) k. \tag{A.83}$$

We have used an exponential function to model the effect of damage S on strain-rate, and it is noted that in Schapery (1981) a discussion is made of the effect of damage on the complementary energy function Φ_f^R:

$$\Phi_f^R = \Phi_{f0}^R \exp(\lambda S),$$
(A.84)

where Φ_{f0}^R is the "effective energy" without damage and λ is a constant. Figure A.4 illustrates the stepwise increase in damage.

Our initial formulations were based on models for crack formation in which we used relationships for the crack density N which involved expressions of the kind $\int_0^t \sigma^q d\tau$, as described in Sections 6.10 and 8.1. This fits naturally with the theory outlined above. We found the need to deal with hydrostatic pressure in the formulation, and several empirically based features were added. The final relationships were of the kind

$$S = \int_0^t \left[f_1(p) \left(\frac{s}{s_0}\right)^{q_1} + f_2(p) \left(\frac{s}{s_0}\right)^{q_2} \right] d\tau,$$
(A.85)

with terms for low and high pressure, the latter dealing with recrystallization. See, for example, equation (8.4) et seq. There is ample ground for further exploration of the basis for the modelling of damage processes in ice in the preceding analysis.

Appendix B Preparation of Laboratory Ice Test Samples

It was considered important to obtain results that were reproducible by other workers, so that standard procedures were decided upon. It was also considered necessary to prepare specimens for uniaxial and triaxial testing, as well as in indentation testing, in such a way that they captured the main physical manifestations that had been observed, and were not dominated by local inhomogeneities. Ice is very prone to localize its failure where there are irregularities, for example in crystal size, or boundary conditions. The triaxial samples were necessarily small to be used in the triaxial cell, so that we aimed at isotropic, granular ice with a relatively small and uniform grain size. It should be emphasized that the actual physical condition as observed in the field showed formation of high-pressure zones with extreme localization dominated by stress and strain gradients. The result was the relatively uniform texture of the deformed material observed in the layers. Local inhomogeneities did not play a role in the processes.

The following summarizes the main features of speciment preparation. There were small differences in detail of procedures between some of the investigators.

B.1   Triaxial Test Specimens

All tests were conducted using laboratory prepared granular ice following the procedure detailed by Stone et al. (1989), Melanson et al. (1999) and Barrette and Jordaan (2003). Large blocks of bubble-free ice were either produced in the laboratory or purchased, and then crushed using a commercial ice crusher. The crushed ice was sieved using mesh between 2.00 mm and 3.36 mm sizes. The seed ice was vacuum moulded using a Fiberglas mould. First, the mould was filled with seed ice and then sealed using a rubber membrane. Next, a vacuum was applied in order to evacuate air from the inter-seed voids. Distilled, deaerated water with a temperature of approximately $0°C$ was then drawn into the mould under a vacuum. The ice was allowed to freeze at $-10°C$ for at least 3 days.

Four cylindrical samples 155 ± 0.2 mm in length and 70 ± 0.05 mm in diameter were machined from each granular ice block. Blocks of ice adjacent to the ends of the cylinders were stored for later thin-section analysis of the undamaged material. The machined samples were stored at $-15°C$ until the day prior to testing. (Some minor variations of these dimensions were used in some of the test series.) An average grain diameter of 3 mm and random c-axis orientation were achieved. These results were checked experimentally—see Stone et al. (1989).

Samples were allowed to equilibrate for at least 12 hours at $-10°C$ prior to being tested with a Materials Testing Systems (MTS) test frame fitted with a Structural Behaviour Engineering Laboratories Model 10 triaxial cell. The MTS system consisted or two servo-controlled hydraulic rams that applied axial load and confining pressure independently. The rams were controlled using MTS TestStar IT software, operated on an OS/2-based 486 computer. The computer and software also performed data acquisition for each test. A prepared sample was mounted on hardened-steel end platens of the same diameter, and the assembly was then enclosed in a latex jacket to exclude the confining medium (silicone oil). The specimen assembly was then placed inside the triaxial cell, and (in the early tests) two linear variable differential transformers (LVDTs) were clamped to the sides of the jacketed specimen. In the tests to low strain levels, the LVDTs were used to monitor axial displacement and to control the advance rate of the ram. For tests to high strain levels, large sample distortions prevented the use of LVDTs. In these tests, the displacement of the ram was used to control the ram advance rate directly. The confining cell was closed and filled with silicone oil, and a small axial load was applied at ambient pressure to ensure contact between the specimen and piston. The confining pressure was then increased at 0.5 MPa s^{-1}, and the axial deformation applied to the sample. At the end of each test the axial load was quickly removed and the confining pressure was then released gradually. The sample was removed from the cell immediately after testing and inspected for jacket leaks and large-scale fracture. When immediate thin sectioning was not possible, samples were stored at $-30°C$ to inhibit grain growth and other microstructural change until thin sections could be prepared and photographed.

B.2   Laboratory Indentation Tests

Barrette et al. (2002) initiated the programme which was followed up by Wells et al. (2011) and others. Large blocks of bubble-free ice were purchased commercially and then crushed using a commercial ice crusher. The crushed ice was sieved to obtain seed ice with sizes in the range of either 2.00 to 3.35 mm or 3.75 to 4.25 mm. The seed ice was vacuum moulded using a Fiberglas mould. First, the mould was filled with seed ice and then sealed using a rubber membrane. Next, a vacuum was applied in order to evacuate air from the inter-seed voids. Distilled, deaerated water with a temperature of approximately 0°C was then drawn into the mould under a vacuum. The ice was allowed to freeze at $-10°C$ for at least 3 days. Finally, the mould was then released at room temperature and the moulded specimens were returned to approximately $-18°C$ for storage. A few days prior to testing, the specimens were moved to the test room, which was held at the test temperature of $-10°C$. The finished test specimens of bubble-free, polycrystalline ice had an average grain size of about 4 mm and overall dimensions of $20 \times 20 \times 10$ cm.

Some tests (Barrette and O'Rourke) were conducted on specimens of ice contained within a steel mould. O'Rourke et al. (2016a) state that each test was conducted at −10°C on an ice sample grown and confined in a cylindrical steel pipe section with an inner diameter of ~154 mm, wall thickness of ~7 mm and 150 mm in length. Seed ice sifted between mesh sizes of 2.00 and 3.35 mm at a temperature of about −13°C, along with distilled water at just above freezing, were used to fill the steel moulds, which were then left to freeze at −13°C. Once frozen, the moulds were allowed to equilibrate to −10°C for at least 12 hours before testing. Prior to testing, the bottom of each mould was milled to remove excess ice from the freezing process, then melted flush to the edge of the steel cylinder, and then clamped to a flat, steel platen attached to the MTS actuator. Each test was performed by pressing the ice at a constant actuator displacement rate into the indentor, thus replicating to some extent the process of ice moving at a constant far-field drift speed into a fixed, compliant structure. Thin sectioning was performed on essentially all samples after indentation by cutting a vertical section from the indented region and using the double-microtome technique presented by Sinha (1977).

References

Adams, J. M., Valtonen, V. and Kujala, P. 2019. Validation of the line-like nature of ice-induced loads using an inverse method. *Proc. 25th Int. Conf. on Port and Ocean Eng. Arctic Cond. (POAC)*, Delft.

Akagawa, S., Nakazawa, N. and Sakai, M. 2000. Ice failure mode predominantly producing peak-ice-load observed in continuous ice load records. *Proc. 10th Int. Offshore and Polar Eng. Conf., Seattle, Washington (ISOPE)*, vol. 2, p. 613.

Alaneme, K. K. and Okotete, E. A. 2019. Recrystallization mechanisms and microstructure development in emerging metallic materials: a review. *J. of Sci. Adv. Mater. Devices.* 4: 19–33.

Anderson, T. L. 2005. *Fracture Mechanics: Fundamentals and Applications.* Taylor & Francis.

Andrade, E. N. da C. 1910. On the viscous flow in metals, and allied phenomena. *Proc. Roy. Soc. Lond., A.* 84:1–12.

Andrews, R. M. 1985. Measurement of the fracture toughness of glacier ice. *J. Glaciol.* 31(108):171–176.

Ashby, M. F. and Duval, P. 1985. The creep of polycrystalline ice. *Cold Reg. Sci. Technol.* 11(3):285–300.

Ashby, M. F., Palmer, A. C., Thouless, M., Goodman, D. J., Howard, M. W., Hallam, S. D., Murrell, S. A. F., Jones, N., Sanderson, T. J. O., Ponter, A. R. S., 1986. Non-simultaneous failure and ice loads on arctic structures. *Proc. OTC 1986.* Houston. Paper No. OTC 5127, pp. 399–404.

Atkins, A. G. and Caddell, R. M. 1974. The laws of similitude and crack propagation. *Int. J. Mech. Sci.* 16:541–548.

Atkins, A. G. and Mai, Y. W. 1985. *Elastic and Plastic Fracture.* Ellis Horwood.

Barenblatt, G. I. 1962. The mathematical theory of equilibrium cracks in brittle fracture. *Adv. Appl. Mech.* VII:55–129.

Barnes, P., Tabor, D. and Walker, J. C. F. 1971. The friction and creep of polycrystalline ice. *Proc. Roy. Soc. Lond. A.* 324:127–155.

Barrette, P. D. 2001. Triaxial testing of ice: a survey of previous investigations. *Proc. Conf. Port Ocean Eng. Arct. Cond. (POAC)*, Ottawa, vol. 3, pp. 1375–1384.

Barrette, P. D. 2014. *Ice-Structure Interaction and High-Pressure Zones: Analysis of Experimental Data and Constitutive Modeling of Ice.* Draft Report to Memorial University, St. John's, Canada.

Barrette, P. D. and Jordaan, I. J. 2001a. Creep of ice and microstructural changes under confining pressure. *IUTAM Symp. Creep Struct.,* S. Murakami and N. Ohno (Eds.), pp. 479–488. Kluwer Academic Publishers.

Barrette, P. D. and Jordaan, I. J. 2001b. *Compressive Behaviour of Confined Polycrystalline Ice.* PERD/CHC Report 4–77. Report to National Research Council Canada. December.

Barrette, P. D. and Jordaan, I. J., 2002. Can dynamic recrystallization and bulk pressure melting explain characteristics of ice crushing? Discussion. Ice in the Environment, *Proc. 16th Int. Symp. Ice, IAHR*, Dunedin, New Zealand vol. 3, pp. 335–354.

Barrette, P. D. and Jordaan, I. J. 2003. Pressure-temperature effects on the compressive behavior of laboratory-grown and iceberg ice. *Cold Reg. Sci. Technol.* March, 36(1–3):25–36.

Barrette, P. D., Pond, J. and Jordaan, I. J. 2002. Ice damage and layer formation in small scale indentation experiments. Ice in the Environment, *Proc. 16th Int. Symp. Ice, IAHR*, Dunedin, New Zealand, vol. 3, pp. 246–253.

Barrette, P. D., Pond, J., Li, C. and Jordaan, I. J. 2003. *Laboratory-Scale Indentation of Ice.* Technical Report for The National Research Council, Program on Energy Research and Development (PERD), PERD/CHC Report 4–81.

Bazant, Z. P. and Planas, J. 1998. *Fracture and Size Effect in Concrete and Other Quasibrittle Materials.* CRC Press.

Biot, M. A. 1954. Theory of stress-strain relations in anisotropic viscoelasticity and relaxation phenomena. *J. Appl. Phys.* 15(11): 1385–1391.

Biot, M. A. 1958. Linear thermodynamics and the mechanics of solids. *Proc. 3rd. U. S. Nat. Cong. Appl. Mech. ASME*, Brown University, Providence, R.I. pp. 1–18.

Blenkarn, K. A. 1970. Measurement and analysis of ice forces on Cook Inlet structures. *Proc. OTC*, Paper 1261.

Blount, H., Glen, I. F., Cornfort, G. and Tarn, G. 1981. *Results of Full Scale Measurements Aboard CCGS Louis St. Laurent during a 1980 Fall Arctic Probe.* Report for Canadian Coast Guard by Arctec Canada Ltd., vols. I and II.

Bolotin, V. V. 1969. *Statistical Methods in Structural Mechanics.* Holden-Day. Translated from the Russian by Samuel Aroni.

Bridgman, P. W. 1912. Water, in the liquid and five solid forms, under pressure. *Am. Acad. Arts Sci.*, Proc. 47:441–558.

Bridgman, P. W. 1941. *The Nature of Thermodynamics.* Harvard University Press.

Brill, R. and Camp, P. R. 1961. *Properties of Ice.* Technical Report, U. S. Army Cold Regions Research and Engineering Laboratory.

Broek, D. 1986. *Elementary Engineering Fracture Mechanics.* Martinus Nijhoff Publishers.

Brown, P. W., Jordaan, I. J., Nessim, M. A. and Haddara, M. M. R. 1996. Optimization of bow plating for icebreakers. *J. Ship Res.* 40(1):70–78.

Browne, T. 2012. Analysis of compressive ice failure during ice-structure interaction. MEng thesis. Memorial University of Newfoundland. St. John's, Canada.

Browne, T., Taylor, R. S., Jordaan, I. and Gürtner, A. 2013. Small-scale ice indentation tests with variable structural compliance. *Cold Reg. Sci. Technol.* 88(2013):2–9.

Budiansky, B. and O'Connell, R. J. 1976. Elastic moduli of a cracked solid. *Int. J. Solids Struct.* 12:81–97.

Budiansky, B., Hutchinson, J. W. and Slutsky, S. 1982. Void growth and collapse in viscous solids. *Mechanics of Solids*, H. G. Hopkins and M. J. Sewell (Eds.), Rodney Hill 60th anniversary volume, Pergamon Press, 13–45.

Butkov, E. 1968. *Mathematical Physics.* Addison-Wesley.

Carter, J. E., Frederking R. M. W., Jordaan, I. J. (Chair), Milne, W. J., Nessim, M. A. and Brown, P. W. 1992. *Review and Verification of Proposals for the Revision of the Arctic Shipping Pollution Prevention Regulations.* Memorial University of Newfoundland, Ocean Engineering Research Centre, report submitted to Canadian Coast Guard, Arctic Ship Safety.

Carter, J., Daley, C., Fuglem, M., Jordaan, I., Keinonen, A., Revill, C., Butler, T., Muggeridge, K. and Zou, B. 1996. *Maximum Bow Force for Arctic Shipping Pollution Prevention*

Regulations. Phase II. Transport Canada Publication No. TP 12652, Transport Canada File No. AMNS 8103-100-8-5, Unclassified: January 1996.

C-CORE. 2017. *Validation of Pressure-Area Scale Effect and Extreme Design Forces Using Full Scale Measured Ship Ram Data.* C-CORE Report R-14-091-1101, Revision 2 FINAL.

Chen, A. C. T. and Lee, J. 1986. Large scale ice strength tests at slow strain rates. *Proc. 5th Int. Offshore Mech. Arctic Eng. Symp. (OMAE),* Tokyo, Japan, 374–378.

Christensen, R. M. 1971. *Theory of Viscoelasticity:An Introduction.* Academic Press.

Cocks, A. C. F. 1989. Inelastic deformation of porous materials. *J. Mech. Phys. Solids.* 37: 693–715.

Cole, D. M. 1986. Effect of grain size on the internal fracturing of polycrystalline ice. CRREL Report 86–5.

Cottrell, A. H. 1964. *The Mechanical Properties of Matter.* Wiley.

Cottrell, A. 1975. *An Introduction to Metallurgy.* Second Edition, Edward Arnold.

Croasdale, K. R. and Marcellus, R. W. 1981. Ice forces on large marine structures. *IAHR Ice Symp.,* Quebec City.

Croasdale, K. R. 1984. The limiting driving force approach to ice loads. *Proceedings OTC.* Houston. OTC Paper 4716.

Croasdale, K., Frederking, R., Jordaan, I. and Noble, P. 2016. *Engineering in Canada's Northern Oceans: Research and Strategies for Development.* Canadian Academy of Engineering. ISBN: 978-1-928194-02-6.

Daley, C., St. John, J. W., and Seibold, F. 1984. Analysis of extreme ice loads measured on USCGC polar sea. *Trans. SNAME,* New York, November, 1984.

Daley, C., St. John, J. W., Brown, R. and Glen, I. 1986. *Consolidation of Local Ice Impact Pressures Measured Aboard USCGC Polar Sea (1982–1984).* Report prepared for Transport Canada, report No. TP 8533E.

Dempsey, J. P. 1996. Scale effects on the fracture of ice. *Minerals, Met. Mat. Soc.,* Arsenault, R. J., Cole, D., Gross, T., Kostroz, G., Liaw, P. K., Parameswaran, S., and Sizek, H. Eds., The Johannes Weertman Symposium, 351–361.

Dempsey J. P., DeFranco S. J., Adamson R. M. and Mulmule S. V. 1999. Scale effects on the in-situ tensile strength and fracture of ice. I: Large-grained freshwater ice at Spray Lakes reservoir, Alberta. II: First-year sea ice at resolute, NWT. *Int J. Fract.,* I, vol. 95, 1999, pp. 325–45;II 347–66.

Dempsey, J. P., Palmer, A. C. and Sodhi, D. S. 2001. High pressure zone formation during compressive ice failure. *Eng. Fract. Mech.* 68: 1961–1974.

Dieter, G. E. 1986. *Mechanical Metallurgy.* McGraw-Hill.

Dome Petroleum. 1982. *Full Scale Measurements of the Ice Impact Loads and Response of the Canmar Kigoriak – August and October, 1981.* Prepared by Dome Petroleum Limited.

Duthinh, D. 1992. Pressure of crushed ice as Mohr-Coulomb material against flat, axisymmetric indentor. *J. Cold Regions Eng., ASCE.* 6, 4:139–151.

Duva, J. M. and Hutchinson, J. W. 1984. Constitutive potentials for dilutely voided nonlinear materials. *Mech. Mater.* 3, 41–54.

Duval, P., Ashby, M. F. and Anderman, I. 1983. Rate-controlling processes in the creep of polycrystalline ice. *J. Phys. Cem.* 87:4066–4074.

Duval, P., Kalifa, P. and Meyssonnier, J. 1991. Creep constitutive equations for polycrystalline ice and effect of microcracking. Ice–Structure Interaction. *Proc. IUTAM-IAHR Symp. Ice–Structure Interact.,* St. John's, Newfoundland, Canada, August 1989, Springer-Verlag, 1991, 55–66.

Duval, P. and Castelnau, O. 1995. Dynamic recrystallization of ice in polar ice sheets. *J. de Physique III*, Colloque C3, vol. 5, C3-197–C3-205.

Ehlers, S., Cheng, F., Jordaan, I., Kuehnlein, W., Kujala, P., Luo, Y., Oh, Y. T., Riska, K., Sirkar, J., Terai, K. and Valkonen, J. 2017. Towards mission-based structural design for arctic regions. *Ship Technol. Res., (Schiffstechnik)*, vol. 64(3): 115–128.

Eshelby, J. D. 1971. Fracture mechanics. *Sci. Prog.*, Oxf. 59:161–179.

Etheridge M. A. and Wilkie, J. C. 1979. Grain size reduction, grain boundary sliding, and the flow strength of mylonites. *Techtonophysics*, 58:159–178.

Fenz, D., Younan, A., Pierce, G., Barrett, J., Ralph, F. and Jordaan, I. 2018. Field measurement of the reduction in local pressure from ice management. *Cold Reg. Sci. Technol.*, 156:75–87.

Fett, T. 2008. *Stress Intensity Factors – T-Stresses – Weight Functions*. Universitätsverlag Karlsruhe.

Findley, W. N., Lai, J. S., and Onaran, K. 1976. *Creep and Relaxation of Nonlinear Viscoelastic Materials*. North-Holland. Reprinted by Dover in 1989.

Flügge, W. 1967. *Viscoelasticity*. Blaisdell Publishing Company.

Frederking, R. and Gold, L. W. 1975. Experimental study of edge loading of ice plates. *Can. Geotech. J.*, 12(4):456–463.

Frederking, R. M. W., Jordaan, I. J. and McCallum, J. S. 1990. Field tests of ice indentation at medium scale: Hobson's Choice Ice Island, 1989, *Proc. 10th. Int. Symp. Ice, IAHR*, Espoo, Finland, vol. 2, 1990, 931–944.

Frederking, R. 1999. The local pressure-area relation in ship impact with ice. *Proc. 15th Int. Conf. Port Ocean Eng. Arctic Cond., POAC'99*, Helsinki, vol. 2, 687–696.

Frederking, R. M. W. 2004. Ice pressure variations during indentation. *Proc. 17th IAHR Int. Symp. Ice*. 2004. v2. p. 307.

Frederking, R. 2005. Local ice pressures on the Oden 1991 polar voyage. *Proc. 18th Int. Conf. Port Ocean Eng. Arctic Cond., POAC05*, vol. 1, 353–363, Potsdam, NY, USA.

Frederking, R. and Kubat, I. 2007. A Comparison of Local Ice Pressure and Line Load Distributions from Ships Studied in the SAFEICE Project. *Proc. 10th International Symposium on Practical Design of Ships and Other Floating Structures*. Houston, Texas, United States of America.

Frederking, R., Sudom, D., Bruce, J., Fuglem, M., Jordaan, I. and Hewitt. K. 2010. *Analysis of Multi-year Ice Loads Molikpaq Data (1985–86 winter deployment)*. Technical Report CHC-CTR-068 January.

Frederking, R., Hewitt, K., Jordaan, I., Sudom, D., Bruce, J., Fuglem, M. and Taylor, R. 2011. Overview of the Molikpaq multi-year ice load analysis. *Proc. 21st Int. Conf. Port Ocean Eng. Arctic Cond., POAC11*.

Fuglem, M., Jordaan, I. J. and Bruce, J. 2011. Estimate of the peak impact load on the Molikpaq by a multi-year floe on May 12, 1986 based on floe deceleration. *Proc. 21st Int. Conf. Port Ocean Eng. Arctic Cond., POAC11*.

Fuglem, M., Parr, G. and Jordaan, I. J. 2013. Plotting positions for fitting distributions and extremal analysis. *Can. J. Civ. Eng.* 40: 130–139.

Fuglem, M., Richard, M. and King, T. 2014. An implementation of ISO 19906 formulae for global sea ice loads within a probabilistic framework. *OTC Arctic Technol. Conf.*, Houston, Texas: Offshore Technology Conference, 2014. https://doi.org/10.4043/24548-MS.

Fuglem. M, Stuckey, P. and Tillson, P. 2016. Evaluation of global ice strength for design iceberg impact loads. *OTC Arctic Technol. Conf.*, St. John's, OTC-27383-MS.

Fuglem, M., Stuckey, P., Huang, Y., Barrett, J., Thijssen, J., King, T. and F. Ralph. 2020. Iceberg disconnect criteria for floating production systems. *Safety Extreme Environ.* 2, no. 1 (April 2020): 15–36. https://doi.org/10.1007/s42797-019-00007-4.

Fung, Y. C. 1965. *Foundations of Solid Mechanics*. Prentice-Hall, New Jersey.

Gagnon, R. E. and Sinha, N. K. 1991. Energy dissipation through melting in large scale indentation experiments on multiyear ice. *Proc. 10th OMAE, ASME*, vol. IV, pp 157–161.

Gagnon, R. E. 1994. Generation of melt during crushing experiments on freshwater ice. *Cold Reg. Sci. Technol.* 22, 385–398.

Gagnon, R. 2002. Can dynamic recrystallization and bulk pressure melting explain characteristics of ice crushing? *Proc. 16th Int. Symp. Ice, IAHR*, Dunedin, New Zealand, 2002.

Garlick, S. R. and Gromet, L. P. 2004. Diffusion creep and partial melting in high temperature mylonitic gneisses, Hope Valley shear zone, New England Appalachians, USA. *J. Metamorphic Geol.* 22:45–62.

Glen I. F. and Blount H. 1984. Measurements of ice impact pressures and loads onboard CCGS Louis S. St. Laurent. *Proc. 3rd Offshore Mech. Arctic Eng. (OMAE) Symp.*, vol. III . ASME, New Orleans, LA, 1984. 246–52.

Glen, J. W. 1955. The creep of polycrystalline ice. *Proc. Roy. Soc. Lond., Ser. A*, 228, 519–538.

Gold, L. W. 1963. Crack formation in ice plates by thermal shock. *Can. J. Phys.* 41:1712–28.

Goodman, D. J. and Tabor, D. 1978. Fracture toughness of ice: a preliminary account of some new experiments. *J. Glaciol.* 21(85):651–660.

Graham, G. A. 1968. The correspondence principle of linear viscoelasticity theory for mixed boundary value problems involving time-dependent boundary regions. *Quart. Appl. Math.*, 26:167–174.

Griffith, A. A. 1921. The phenomena of rupture and flow in solids. *Philos. Trans. Roy. Soc. Lond. A: Mathematical, Physical and Engineering Sciences*, 221(582–593):163–198.

Gross, B. 1968. *Mathematical Structure of the Theories of Viscoelasticity*. Hermann, Paris.

Guo, Fengwei. 2012. A spectral model for simulating continuous crushing ice load. *Proc. 21st IAHR Int. Symp. Ice*. Dalian, China, June, 1035–1045.

Haddad, Y. M. 1995. *Viscoelasticity of Engineering Materials*. Chapman Hall.

Hamza, H. and Muggeridge, D. B. 1979. Plane strain fracture toughness (K_{1c}) of freshwater ice. *Proc. 5th Int. Conf. Port Ocean Eng. Arctic Cond., POAC79*:697–707.

Hardy, M. D., Jefferies, M. G., Rogers, B. T. and Wright, B. D. 1996. *Molikpaq Ice Loading Experience*. Report submitted to National Energy Board, Canada PERD/CHC Report 14–62.

Harper, B. D. 1986. A uniaxial nonlinear viscoelastic constitutive relation for ice. *J. Energy Resour. Technol.* June, 108(2): 156–160

Hewitt, K. 2010. *Estimates of Ice Loads on the Molikpaq at Amauligak I-65 Based on Geotechnical Analyses and Responses*. Report January 2010. Available from www.engr.mun.ca/~2007JIPonMolikpaq/.

Hobbs, P. V. 1974. *Ice Physics*. Oxford University Press.

Horii, H. and Nemat-Nasser, S. 1983. Overall moduli of solids with microcracks: load-induced anisotropy. *J. Mech. Phys. Solids*, 31(2):155–171.

IACS. 2006. *Structural Requirements for Polar Class Ships – Technical Background*. IACS UR I2, AHG/PSR, August 2006.

IACS. 2011. *Requirements Concerning Polar Class, International Association of Classification Societies*. IACS Req. 2011.

Ian Jordaan and Associates. 2007. *Analysis of JOIA Data: Probabilistic Averaging For Global Load Estimation.* Final Report Submitted to Agip KCO, Conoco Phillips and Shell.

Ian Jordaan and Associates. 2010. *Investigation of Molikpaq 1986 Ice Loading Events and Evaluation of Load Measuring Devices.* Report to Canadian Hydraulics Centre, National Research Council of Canada, January. Available in www.engr.mun.ca/~2007 JIPonMolikpaq/.

IDA. 2007. Jordaan, I. J., Li, C., Mackey, T., Stuckey, P. and Sudom, D. *Ice Data Analysis and Mechanics for Design Load Estimation.* (IDA project). Final Report. Prepared for NSERC, C-CORE, Chevron Canada Resources, National Research Council of Canada, Petro-Canada, Husky Energy.

Inglis, C. E. 1913. Stresses in plates due to the presence of cracks and sharp corners. *Trans. Inst. Nav. Archit.*, 55:219–241.

ISO 19906. 2019. *Petroleum and Natural Gas Industries–Arctic Offshore Structures.*

Jacka, T. H. and Maccagnan, M. 1984. Ice crystallographic and strain rate changes with strain in compression and extension. *Cold Reg. Sci. Technol.*, 8(3):269–286.

Johari G. P., Pasheto W. and Jones S. J., 1994. Intergranular liquid in solids and premelting of ice. *J. Chem. Phys.* 100(6):4548–4553.

Jonas, J. J. and Müller, F., 1969. Deformation of ice under high internal shear stresses. *Can. J. Earth Sci.*, vol. 6; 963–967

Jones S. J. 1978. Triaxial testing of polycrystalline of ice. *Proc. 3rd Int. Conf. Permafrost*, Edmonton, vol. 1:671–674.

Jones S. J. 1982. The confined compressive strength of polycrystalline ice. *J. Glaciol.* 28: 171–177.

Jordaan, I. J., Nessim, M. A., Ghoneim, G. A. and Murray, A. M. 1987. A rational approach to the development of probabilistic design criteria for arctic shipping. *Proc. 6th Int. Offshore Mech. Arctic Eng. Symp. (OMAE),* Houston, vol. IV:401–406.

Jordaan, I. J. and McKenna, R. F. 1988. Constitutive relations for the creep of ice. *Proc. 9th Int. Symp. Ice, IAHR*, Sapporo, Japan, vol. 3: 47–58.

Jordaan I. J. and Timco G. W. 1988. The dynamics of the ice crushing process. *J. Glaciol.* 34(118):318–26.

Jordaan, I. J. and McKenna, R. F. 1991. Processes of deformation and fracture of ice in compression. *Ice–Structure Interaction. Proc. IUTAM-IAHR Symp. Ice-Structure Interaction*, St. John's, Newfoundland, Canada, August 1989, Springer-Verlag, 1991:283–309.

Jordaan, I. J. and McKenna, R. F. 1988. Modelling of progressive damage in ice. *Proc. 9th Int. Symp. Ice, IAHR*, Sapporo, Japan, vol. 2:585–624. Also in 4th. State-of-the-Art Report (G. Timco, Ed.) Working Group on Ice Forces, CRREL Special Report 89–5, 1989, 125–165.

Jordaan, I. J. and Xiao, J. 1992. Interplay between damage and fracture in ice-structure interaction. *Proc. 11th IAHR Int. Ice Symp.*, Banff, Alberta, vol. 3:1448–1467.

Jordaan, I. J., Stone, B. M., McKenna, R. F. and Fuglem, M. K. 1992. Effect of microcracking on the deformation of ice. Published in *Proc. 43rd. Can. Geotech. Conf,*, Université Laval, Québec, vol. 1, 1990, pp. 387–393; *Can. Geotech. J.*, 29:143–150.

Jordaan, I. J., Maes, M. A., Brown, P. W. and Hermans, I. P. 1993a. Probabilistic analysis of local ice pressures. *Proc. 11th IAHR Int. Ice Symp,* Banff, Alberta, vol. II, 1992, 7–13; Also in *J. Offshore Mech. Arctic Eng.*,115(1):83–89.

Jordaan, I. J., Xiao, J. and Zou, B. 1993b. Fracture and damage of ice: towards practical implementation. *First Joint ASCE-EMD, ASME-AMD, SES Meeting*, Virginia, June. ASME, AMD-vol. 163, 1993, 251–260.

Jordaan, I. J. and Singh, S. K. 1994. Compressive ice failure: critical zones of high pressure. *Proc. 12th Int. IAHR Ice Symp.*, Trondheim Norway, vol. 1:505–514.

Jordaan, I. J., Fuglem M. and Matskevitch, D. G. 1996. Pressure-area relationships and the calculation of global ice forces. *Proc. 13th IAHR Int. Symp. Ice*, Beijing, China, vol. 1, 166–175.

Jordaan, I. J., Matskevitch, D. M. and Meglis, I. 1999. Disintegration of ice under fast compressive loading. *Proc. Symp. "Inelasticity and Damage in Solids subject to Microstructural Change"*, Memorial University of Newfoundland, 1997: 211–231. Extended version: *Int. J. Fract.* 97(1–4):279–300.

Jordaan, I. J. 2001. Mechanics of ice-structure interaction. *Eng. Fract. Mech.*, 2001; 68:1923–1960.

Jordaan, I. J. and Pond, J. 2001. Scale effects and randomness in the estimation of compressive ice loads. Proc. IUTAM Symp. Scaling Laws in Ice Mech. Dyn., Fairbanks, Alaska, 2000. Ed. J. P. Dempsey and H. H. Shen, Kluwer, 2001, 43–54.

Jordaan, I. J. 2005. *Decisions under Uncertainty: Probabilistic Analysis for Engineering Decisions.* Cambridge University Press, 2005, 672.

Jordaan, I., Li, C., Sudom, D., Stuckey, P. and Ralph, F. 2005a. Principles for local and global ice design using pressure- area relationships. *Proc. 18th Int. Conf. Port Ocean Eng. Arctic Cond. (POAC'05)*, Potsdam, N. Y., vol. 1:375–385.

Jordaan, I., Li, C., Barrette, P., Duval, P. and Meyssonnier, J. 2005b. Mechanisms of ice softening under high pressure and shear. *Proc. 18th Int. Conf. Port Ocean Eng. Arctic Cond. (POAC'05)*, Potsdam, N. Y., vol. 1:249–259.

Jordaan, I., Frederking, R. and Li, C. 2006. Mechanics of ice compressive failure, probabilistic averaging and design load estimation. *Proc. 18th Int. Symp. Ice, IAHR*, Sapporo, Japan, vol. 1:223–230.

Jordaan, I. J., Li, C., Mackey, T., Stuckey, P. and Sudom, D. 2007a. *Ice Data Analysis and Mechanics for Design Load Estimation.* (IDA project), Final Report. Prepared for NSERC, C-CORE, Chevron Canada Resources, National Research Council of Canada, Petro-Canada, Husky Energy.

Jordaan, I., Taylor, R. and Reid, S. 2007b. Fracture, Probabilistic averaging and the scale effect in ice-structure interaction. *Proc. 19th Int. Conf. Port Ocean Eng. Arctic Cond., POAC'07.* Dalian, China, vol. 1:296–304.

Jordaan, I. J., Taylor, R. S. and Wells, J. 2009. Ice crushing, damaged layers, and pressure-area relationships. *Proc. 20th Int. Conf. Port Ocean Eng. Arctic Cond., POAC'09.* Paper 09–128.

Jordaan, I. J., Bruce, J., Masterson, D. and Frederking, R. M. W. 2010. Local ice pressures for multiyear ice accounting for exposure. *Cold Reg. Sci. Technol.*, 61: 97–106.

Jordaan, I. 2010. Viscoelasticity and localization in compressive ice failure, indenter tests, analysis of crushed layer, and vibrations. Background Information Paper for Workshop on Ice-induced Vibrations, Oslo, Norway, November 19 & 20.

Jordaan, I. J. and Bruce, J. 2010. Notes on local ice pressures and velocity effect. Internal Report, C-CORE.

Jordaan, I., Bruce, J., Hewitt, K. and Frederking, R. 2011. Re-evaluation of ice loads and pressures measured in 1986 on the Molikpaq structure. *Proc. 21st Int. Conf. Port Ocean Eng. Arctic Cond. (POAC'11)*. July 10–14, Montréal, Canada.Paper 11–130.

Jordaan, I., Stuckey, P., Bruce, J., Croasdale, K. and Verlaan, P. 2011. Probabilistic modeling of the ice environment in the NE Caspian Sea and associated structural loads. *Proc. 21st Int. Conf. Port Ocean Eng. Arctic Cond. (POAC'11)*. July 10–14, Montréal, Canada. Paper POAC11-133.

Jordaan, I. and Gosine, P. 2012. The Titanic disaster and ice mechanics: completing the picture. *Proc. IceTech 2012*, Banff, Alberta, September, Paper ICETECH12-149-R0.E

Jordaan, I., Taylor, R. and Derradji-Aouat, A. 2012. Scaling of flexural and compressive ice failure. *Proc. 31st Int. Conf. Ocean Offshore Arct. Eng., OMAE 2012*, July, Rio de Janeiro, Paper OMAE2012-84033.

Jordaan I. and Barrette P. 2014. Mechanics of dynamic ice failure against vertical structures. *Proc. 33rd Int. Conf. Ocean Offshore Arct. Eng., OMAE 2014*, June 8–13, San Francisco. ASME. Paper OMAE2014-24406.

Jordaan, I., Stuckey, P., Liferov, P. and Ralph, F. 2014. Global and local iceberg loads for an arctic floater. *Proc. Arct. Technol. Conf.*, Houston, U. S. A.: OTC 24628, 2014. https://doi.org/10.4043/24628-MS.

Jordaan, I. 2015. Some issues in ice mechanics. *Proc. 34th Int. Conf. Ocean Offshore Arct. Eng., OMAE 2015*. Paper No. OMAE2015-42042.

Jordaan, I., O'Rourke, B., Turner, J., Moore, P. and Ralph, F. 2016. Estimation of ice loads using mechanics of ice failure in compression. *Proc. Arct. Technol. Conf.*, St John's, Newfoundland, paper OTC 2738.

Jordaan, I., Hewitt, K. and Frederking, R. 2018. Re-evaluation of ice loads on the Molikpaq structure measured during the 1985–86 season. *Can. J. Civ. Eng.* 45: 153–166.

Kachanov, L. M. 1958. On rupture time under condition of creep. *Izvestia Akademi Nauk SSSR, Otd. Tekhn. Nauk*, No. 8, pp. 26–31 (in Russian).

Kachanov, M. 1993. Elastic solids with many cracks and related problems. *Adv. Appl. Mech.*, vol. 30, Academic Press.

Kachanov, M., Tsukrov, I. and Shafiro, B. 1994. Effective moduli of solids with cavities of various shapes. *Appl. Mech. Rev.* 47, Part 2, S151–S174.

Kalifa, P., Ouillon, G. and Duval, P. 1992. Microcracking and the failure of polycrystalline ice under triaxial compression. *J. Glaciol.*, 38(128), 65–76.

Kärnä, T. and Turunen, R., 1990. A straightforward technique for analyzing structural response to dynamic ice action. *Proc. 9th Int. Conf. Offshore Mech. Arct. Eng., OMAE*, Houston, Texas, February 18–23, 1990. ASME vol. 4, 135–142.

Kärnä, T. and Yan, Qu. 2006. *Analysis of the Size Effect in Ice Crushing – edition 2*. Internal report, Technical Research Centre for Finland, February 2006.

Kärnä, T. 2007. Research problems related to time-varying ice actions. *Proc. 19th Int. Conf. Port Ocean Eng. Arct. Cond., POAC'07.*, Dalian, China.

Karr, D. G. and Choi, K. 1989. A three-dimensional constitutive damage model for polycrystalline ice. *Mech. Mater.* 8,1:55–66.

Kasap, S. O. 2006. *Principles of Electronic Materials and Devices*. McGraw-Hill.

Kavanagh, M., O'Rourke, B., Jordaan, I. and Taylor, R. 2015. Observations on the time-dependent fracture of ice. In Ian Jordaan Honoring Symposium on Ice Engineering. *Proc. 34th Int. Conf. Ocean Offshore Arct. Eng., OMAE 2015*. Paper No. OMAE2015-42023. St. John's, Newfoundland, Canada. ASME.

Kavanagh, M. B. 2018. Time-Dependent Aspects of Fracture in Ice. PhD thesis, Memorial University of Newfoundland.

Kavanagh, M. and Jordaan, I. 2022. Time-Dependent Fracture of Ice. *Eng. Fract. Mech.*, Vol. 276, Part A, Paper 108850.

Kendall, K. 1978. Complexities of compression failure. *Proc. Roy. Soc. Lond.* A361: 245–263.

Kennedy K. P., Jordaan I. J., Maes M. A. and Prodanovic A. 1994. Dynamic activity in medium-scale ice indentation tests. *Cold Reg. Sci. Technol.* 22:253–67.

Kheisin, D. E. and Cherepanov, N. W. 1970. Change of ice structure in the zone of impact of a solid body against the ice cover surface. *Problemy Artiki I Anarktiki*. Issues 33–35 (A. F. Treshnikov), Israel Program for Scientific Translations (1973) 239–245.

Kheisin, D. E. 1973. Use of probability methods in estimating the manoeuvering qualities of ships in ice. *Ice Navigation Qualities of Ships*. Ed. D. Kheisin and Yu. Popov. CRREL Translation pp. 40–58.

Kheisin, D. E. and Likhomanov, V. A. 1973. An experimental determination of the specific energy of mechanical crushing in ice by impact. *Problemy Artiki I Anarktiki* 41, 55–61.

Kim, E. and Amdahl, J. 2016. Discussion of assumptions behind rule-based ice loads due to crushing. *Ocean Eng.* 119 (2016) 249–261.

Krausz, A. S. and Eyring, H. 1975. *Deformation Kinetics*. Wiley.

Kry, P. R. 1978. A statistical prediction of effective ice crushing stresses on wide structures. *Proc. 5th Int. IAHR Conf.*, Lulea, Sweden, Part 1:33–47.

Kry, P. R. 1980. Implications of structure width for design ice forces. *Physics and Mechanics of Ice*. Proc. *IUTAM Symposium*. Copenhagen, August 6–10, 1979, Technical University of Denmark. pp. 179–193.

Kujala, P. J. 1991. Safety of ice-strengthened ships in the Baltic Sea. London. *Trans. Roy. Inst. Nav. Archit.* 133 (A), pp. 83–94.

Kujala, P. J. 1996. Modelling of nonsimultaneous ice crushing as a Poisson random process. *Int. J. Offshore Polar Eng.* vol. 6, No.2, pp. 138–143.

Kuon, L. G. and Jonas, J. J. 1973. Effect of strain rate and temperature on the microstructure of polycrystalline ice. *Symp. Phys. Chem. Ice*, Ottawa, Canada, 14–18 August 1972, published by the Royal Society of Canada, 1973. pp. 370–376.

Kurdyumov, V. A. and Kheisin, D. E. 1976. Hydrodynamic model of the impact of a solid on ice. Translated from *Prikladnaya Mekhanika*, 1976;12(10):103–109.

Lakes, R. 2009. *Viscoelastic Materials*. Cambridge.

Lakes, R. S. and Vanderby, R. 1999. Interrelation of creep and relaxation: a modeling approach for ligaments. *J. Biomech. Eng.* 121, 612–615, Dec.

Landau L. D. and Lifshitz E. M. 1970. *Theory of Elasticity*. vol. 7. in Course of Theoretical Physics; 2nd English Ed., Pergamon.

Lawn, B. 1993. *Fracture of Brittle Solids*. Second Ed. Cambridge.

LeClair, E. S., Schapery, R. A. and Dempsey, J. P. 1999. A broad-spectrum constitutive modeling technique applied to saline ice. *Int. J. Fract.* 97:209–226.

Leckie, F. A. 1978. The constitutive equation of continuum creep damage mechanics. *Philos. Trans. R. Soc. Lond.* A **288**, 27–47.

Li, C. (Chuanke) 2002. Finite Element Analysis of Ice-structure Interaction with a Viscoelastic Model Coupled with Damage Mechanics. MEng thesis, Memorial University of Newfoundland, 127.

Li, C., Barrette, P. and Jordaan, I. 2004. High-pressure zones at different scales during ice-structure indentation. *Proc. 23rd Int. Conf. Offshore Mech. Arctic Eng., OMAE,* Vancouver, British Columbia.

Li, C., Jordaan, I. and Barrette, P. 2005. Strain localization and fracture of cylindrical ice specimens under confining pressure. *Proc. 18th Int. Conf. Port Ocean Eng. Arctic Cond. (POAC'05)*. Potsdam, N. Y., Proceedings, vol. 1:213–224.

Li, C. (Chuanke) 2007. Probability and Fracture Mechanics Applied to Ice Load Estimation and Associated Mechanics. PhD thesis, Memorial University of Newfoundland, 190.

Li, C., Jordaan, I. J. and Taylor, R. S. 2010, Estimation of local ice pressure using up-crossing rate. *J. Offshore Mech. Arctic Eng.*, vol. 132, (2010)031501–1to6

Liu, H. W. and Miller, K. J. 1979. Fracture toughness of fresh-water ice. *J. Glaciol.*, 22(86): 135–143.

Liu, B. (Bin) 1994. Numerical Modelling of Medium Scale Indentation Tests. M.Eng thesis, Memorial University of Newfoundland, 106.

Määttänen, M. 1983. *Dynamic ice–structure interaction during continuous crushing*. CRREL Report 83–5. U. S. Army Cold Regions Research and Engineering Laboratory.

Määttänen, M. 2001. Numerical simulation of ice-induced vibrations in offshore structures. *Proc. 14th Nordic Seminar Comput. Mech.,* Lund, Sweden. 2001. 13–28.

Mackey, T., Wells, J., Jordaan, I. and Derradji-Aouat, A. 2007. Experiments on the fracture of polycrystalline ice. *Proc. 19th Int. Conf. Port Ocean Eng. Arctic Cond. (POAC'07).* Dalian, China, 339–349.

Maes, M. A. 1992. Probabilistic behaviour of a Poisson field of flaws in ice subjected to indentation. *Proc. 11th Int. Symp. Ice, IAHR*, Banff, Alberta, vol. 2 871–882.

Manuel, M., Ralph, F. and Jordaan, I. 2016. Using the Event Maximum Method to Further Analyze Full Scale Local Pressure Data. *Proc. Arctic Tech. Conf.*, St. John's, Newfoundland and Labrador, Canada, Paper No. OTC – 27368.

Masterson, D. and Spencer, P., 1992. *Reduction and Analysis of 1990 and 1989 Hobson's Choice Ice Island Indentation Tests Data.* Report prepared by Sandwell Inc.

Masterson D. M., Nevel D. E., Johnson R. C., Kenny J. J. and Spencer P. A. 1992. The medium scale iceberg impact test program. *Proc. 11th Int. Symp. Ice, IAHR*, Banff, Alberta. Banff, Alberta.

Masterson, D. M. and Frederking, R. M. W. 1993. Local contact pressures in ship-ice and structure-ice interactions. *Cold Reg. Sci. Technol.* 21:169–185.

Masterson, D. M., Frederking, R. M. W., Jordaan I. J. and Spencer, P. A. 1993. Description of multiyear ice indentation tests at Hobson's Choice ice island – 1990. *Proc. 12th Int. Conf. Offshore Mech. Arctic Eng., OMAE 1993.* Glasgow. vol. 4:145–55.

Masterson D. M., Spencer P. A., Nevel D. E. and Nordgren R. P. 1999. Velocity effects from multiyear ice tests. *Proc. 18th Int. Conf. Offshore Mech. Arctic Eng., OMAE99,* St. John's, Canada, Paper OMAE99-1127.

Masterson, D. M., Frederking, R. M. W., Wright, B., Karna, T. and Maddock, W. P. 2007. A revised ice pressure–area curve. *Proc. 19th Int. Conf. Port Ocean Eng. Arctic Cond. POAC'07*, Dalian, China. June 27–30.

Masterson, Dan. 2018. *The Story of Offshore Arctic Engineering.* Cambridge Scholars Publishing.

Matskevitch, D. G. and Jordaan, I. J. 1996. Spatial and temporal variations of local ice pressures. *Proc. 15th. Int. Conf. Offshore Mech. Arctic Eng. (OMAE),* June 16–20, Florence, Italy.

McCarty, J. H. and Foecke, T. 2008. *What Really Sank the Titanic.* Citadel Press.

McKenna, R. F., Jordaan, I. J. and Xiao, J. 1990. Damage and energy flow in the crushed layer during rapid ice loading. *Proc. IAHR Symp. Ice*, Espoo, Finland, vol. 3, 1990, pp. 231–245.

Meaney, R., Jordaan, I. J. and Xiao, J. (1996). Analysis of medium scale ice-indentation tests. *Cold Reg. Sci. Technol.* 24, 279–287.

Meglis, I., Melanson, P. and Jordaan I. J. 1999. Microstructural change in ice: II. Creep behavior under triaxial stress conditions, *J. Glaciol.* 45(151):438–448 Colour Plates 423–437.

Melanson, P. M. 1998. Damage and Microstructural Change in Laboratory Grown Ice under High Pressure Zone Conditions. M.Eng. thesis, Memorial University of Newfoundland.

Melanson, P., Meglis, I., Jordaan, I. J. and Stone, B. S. 1999. Microstructural change in ice: I. Constant deformation-rate tests under triaxial stress conditions, *J. Glaciol.* 45(151):417–422 Colour Plates 423–437.

Mellor, M. and Cole, D. M. 1982. Deformation and failure of ice under constant stress or constant strain rate. *Cold Reg. Sci. Technol.* 5:201–219

Mellor, M. and Testa R. 1969. Effect of temperature on the creep of ice. *J. Glaciol.* 8: 131–145.

Metge, M. 1994. Hans Island revisited. *Proc. IAHR Symp. Ice*, vol. 3, 1003–1017, Trondheim, Norway.

Meyssonnier, J. and Duval, P. 1989. Creep behaviour of damaged ice under uniaxial compression: a preliminary study. *Proc. 10th Int. Conf. Port Ocean Eng. Arctic Cond. POAC'89*, Lulea, Sweden. pp. 225–234.

Michel, B. 1978. *Ice Mechanics*. Laval, Québec.

Miller, K. J. 1980. The application of fracture mechanics to ice problems. *Physics and Mechanics of Ice. Proc. IUTAM Symposium.* Copenhagen, 1979.

Muggeridge, K., and Jordaan, I. J. 1999. Microstructural change in ice: III. Observations from an iceberg impact zone, *J. Glaciol.*, 45(151):449–455 Colour Plates 423–437.

Mulmule S. V. and Dempsey J. P. 1998. A viscoelastic fictitious crack model for the fracture of sea ice. *Mech. Time-dependent Mater.*1:331–56.

Nadreau J. P. and Michel B. 1986a. Secondary creep in confined ice samples. *Proc. 8th IAHR Symp. Ice.*, Iowa City, vol. 1, 307–318.

Nadreau J. P. and Michel B. 1986b. Yield envelope for confined ice. *Ice Technology, Proc. 1st Int. Conf.*, 25–36.

Needleman, A. 1988. Material rate dependence and mesh sensitivity in localization problems. *Comput. Methods Appl. Mech. Eng.* vol. 67, pp. 69–85.

Needleman, A. 1991. On the competition between failure and instability in progressively softening solids. *Trans. ASME.* vol. 58:294–298.

Nixon, W. and Schulson, E. 1987. A micromechanical view of the fracture toughness of ice. *J. Phys., Colloques,* 1987, 48 (C1), C1-313–C1-319.

Nixon, W. A. 1988. The effect of notch depth on the fracture toughness of freshwater ice.*Cold Reg. Sci. Technol.* 15, 75–78.

Nordell, B. 1990. Measurement of P-T coexistence curve for ice-water mixture. *Cold Reg. Sci. Technol.* 19:83–88.

Nye J. F. 1991. Rotting of temperate ice. *J. Cryst. Growth*, 113:465–476.

Ochi M. K. 1990. *Applied Probability and Stochastic Processes*. Wiley; 1990.

Offenbacher, E. L., Roselman, I. C. and Tabor, D. 1973. Friction, deformation and recrystallization of single crystals of ice Ih under stress. *Proc. Symp. Phys. Chem. Ice*, Ottawa, Canada, 14–18 August 1972, published by the Royal Society of Canada, 1973.

O'Rourke, B. J, Jordaan, I. J., Taylor, R. S. and Gürtner, A., 2015. Spherical indentation tests on confined ice specimens at small scales. *Proc. 34th Int. Conf. Ocean, Offshore Arctic Eng., OMAE 2015.* Paper No. OMAE2015-42117.

O'Rourke, B. J., Jordaan, I. J., Taylor, R. S. and Gürtner, A., 2016a. Experimental investigation of oscillation of loads in ice high-pressure zones, Part 1: Single indentor system. *Cold Reg. Sci. Technol.* 124 (2016) 25–39.

O'Rourke, B. J., Jordaan, I. J., Taylor, R. S. and Gürtner, A., 2016b. Experimental investigation of oscillation of loads in ice high-pressure zones, Part 2: Double indentor system – coupling and synchronization of high-pressure zones. *Cold Reg. Sci. Technol.* 124 (2016) 11–24.

Palmer A. C., Goodman D. J., Ashby M. F., Evans A. G., Hutchinson J. W. and Ponter A. R. S. 1983. Fracture and its role in determining ice forces on offshore structures. *Ann. Glaciol.* 4:216–21.

Palmer A. C. 2008. *Dimensional Analysis and Intelligent Experimentation.* World Scientific, Singapore.

Palmer, A. and Dempsey, J. 2009. Model tests in ice. *Proc. 20th Int. Conf. Port Ocean Eng. Arctic Cond., POAC09.* Paper POAC09-40.

Palmer, A. C. and Croasdale, K. R. 2013. *Arctic Offshore Engineering.* World Scientific Publishing.

Palmer A. and Bjerkås M. 2013. Synchronization and the transition from intermittent to locked-in ice-induced vibration. *Proc. 22nd Int. Conf. Port and Ocean Eng. Arctic Cond. (POAC),* Espoo, Finland.

Paterson W. S. B., 1994. *The Physics of Glaciers.* Pergamon, Toronto.

Peyton, H. R. 1968. *Sea Ice Forces.* In "Ice Pressures against Structures", National Research Council of Canada, Ottawa, Canada, Technical Memorandum 92, pp. 117–123.

Pikovsky A., Rosenblum M. and Kurths J. 2001. *Synchronization: a Universal Concept in Nonlinear Sciences.* Cambridge University Press.

Ponter, A. R. S., Palmer, A. C., Goodman, D. J., Ashby, M. F., Evans, A. G. and Hutchinson, J. W. 1983. The force exerted by a moving ice sheet on an offshore structure. Part 1. The creep mode. *Cold Reg. Sci. Technol.,* 8:109–118.

Popov, Y., Fadeyev, O., Kheisin, D. and Yakovlev, A. 1967. *Strength of Ships Sailing in Ice.* Sudostroenie Publishing House, Leningrad, 223 (in Russian).

Pounder, E. R. 1965. *The Physics of Ice.* Pergamon.

Rabotnov, Yu. N. 1969. *Creep Problems of Structural Members.* North-Holland, Amsterdam.

Ralph, F., McKenna, R., Crocker,G. and Croasdale, K. 2004. Pressure/area measurements from the Grappling Island iceberg impact experiment. *Proc. 17th IAHR Int. Symp. Ice,* St. Petersburg, Russia. 21–25 June, 171–178.

Ralph, F., Jordaan, I. J. Clark, P. and and Stuckey, P. 2006. Estimating probabilistic iceberg design loads on ships navigating in ice covered waters. *Proc. ICETECH.* Paper No. ICETECH06-05-006, Banff.

Ralph, F. and Jordaan, I. J. 2013. Probabilistic methodology for design of arctic ships. *Proc. 32nd Int. Conf. Ocean Offshore Arctic Eng. (OMAE).* Nantes, France. June 2013.

Ralph, F. E. 2016. Design of Ships and Offshore Structures: A Probabilistic Approach for Multi-Year Ice and Iceberg Impact Loads for Decision-making with Uncertainty. PhD thesis, Memorial University of Newfoundland.

Ralph, F., Jordaan, I. and Manuel, M. 2016. Application of upcrossing rate methodology to local design of icebreaking vessels. *Proc. Arctic Tech. Conf.,* St. John's, NL. Paper no. OTC-27366-MS.

Rice, J. R. 1968. A path independent integral and the approximate analysis of strain concentration by notches and cracks. *J. Appl. Mech.* 35:379–386.

Riska, K. 1987. *On the Mechanics of the Ramming Interaction between a Ship and a Massive Ice Floe.* Technical Research Centre of Finland, Espoo. Publications 43.

Riska, K. 1991 Observations of the line-like nature of ship–ice contact. *Proc. 11th Int. Conf. Port Ocean Eng. Arctic Cond., POAC'91,* St. John's, Canada, 24–28 September, vol. 2, 785–811.

Rist, M. A., Murrell, S. A. F. and Sammonds, P. R. 1988. Experimental results on the failure of polycrystalline ice under triaxial stress conditions. *Proc. 9th IAHR Int. Symp. Ice, IAHR 88,* Sapporo, Japan. Vol.1:118–127.

Rist, M. A., Jones, S. J. and Slade, T. D. 1994. Microcracking and shear fracture in ice. *Ann. Glaciol.* 19:131–137 1.

Ritch, R., St. John, J., Browne, R. and Sheinberg, R. 1999. Ice load impact measurements on the CCGS *Louis S. St. Laurent* during the 1994 Arctic Ocean crossing. *Proc. 18th Int. Conf. Offshore Mech. Arctic Eng. (OMAE)*, July 11–16, 1999, St. John's Newfoundland, paper OMAE99/P&A-1141.

Ritch, R., Frederking, R., Johnston, M., Browne, R. and Ralph, F. 2008. Local ice pressure measured on a strain gauge panel during the CCGS *Terry Fox* bergy bit impact study. *Cold Reg. Sci. Technol.* 52, pp. 29–49.

RMRS. 2014. *Rules for the Classification and Construction of Sea-going Ships.* vol. 1. Russian Maritime Register of Shipping, Saint-Petersburg.

Rogers, B. T., Hardy, M. D., Neth, V. W. and Metge, M. 1986. Performance monitoring of the Molikaq while deployed at Tarsuit P-45. *Proc. Can. Conf. Mar. Geotech. Eng.* 363–383.

Rogers. B, Hardy, M. D., Jefferies, M. G, Wright, B. D. 1998. *DynaMAC: Molikpaq Ice Loading.* Report by Klohn-Crippen, PERD/CHC Report 14–62.

Sanderson, T. J. O. 1988. *Ice Mechanics: Risks to Offshore Structures.* Graham and Trotman.

Sandwell. 1991. *Extensometer Calibration for Ice Load Measurement.* Report 112451, for Gulf Canada Resources Ltd., May 1991.

Savage, S. B., Sayed, M. and Frederking, R. M. W. 1992. Two-dimensional extrusion of crushed ice. Part 2: Analysis. *Cold Reg. Sci. Technol.* 21:37–47.

Sayed, M. and Frederking, R. M. W., 1992. Two-dimensional extrusion of crushed Ice. Part 1: Experimental. *Cold Reg. Sci. Technol.* 21:25–36.

Schapery, R. A. 1964. Application of thermodynamics to thermomechanical, fracture, and birefringent phenomena. *J. Appl. Phys.* 35(5):1451–1465.

Schapery, R. A. 1966. A theory of nonlinear thermoviscoelasticity based on irreversible thermodynamics. *Proc. 5th U. S. Nat. Congr. Appl. Mech.*, ASME, 511–530.

Schapery, R. A. 1968. On a thermodynamic constitutive theory and its application to various nonlinear materials. *Proc. IUTAM Symp.*, East Kilbride, June, Ed. Bruno Boley. Springer. 259–285.

Schapery, R. A. 1969. *Further Development of a Thermodynamic Constitutive Theory: Stress Formulation.* Purdue University Rept. AA&ES 69–2.

Schapery, R. A. 1975a. A theory of crack initiation and growth in viscoelastic media: I. theoretical development. *Int. J. Fract.* 11 (1):141–159.

Schapery, R. A. 1975b. A theory of crack initiation and growth in viscoelastic media II. Approximate methods of analysis. *Int. J. Fract.* 11(3):369–387.

Schapery, R. A. 1975c. A theory of crack initiation and growth in viscoelastic media III. Analysis of continous growth. *Int. J. Fract.* 11(4):549–562.

Schapery R. A. 1981. On viscoelastic deformation and failure behavior of composite materials with distributed flaws. *Adv. Aerosp. Struct. Mat.*, S. S. Wang and W. J. Renton (Eds.), ASME, AD-01, 5–20.

Schapery R. A. 1984a. Time-dependent fracture: continuum aspects of crack growth. *Ency. Mat. Sci. Eng.* M. B. Bever (Ed.), Pergamon Press, Oxford, 1986 pp. 5043–5053.

Schapery, R. A. 1984b. Correspondence principles and a generalized J integral for large deformation and fracture of viscoelastic media. *Int. J. Fract.*, 25:195–223.

Schapery, R. A. 1991a. Models for the deformation behavior of viscoelastic media with distributed damage and their applicability to ice. *Ice–Structure Interaction, Proc. IUTAM/IAHR Symp.* St. John's, Newfoundland, Canada, August 1989, Springer-Verlag, 1991:181–230.

Schapery, R. A. 1991b. Simplifications in the behavior of viscoelastic composites with growing damage. *Inelastic Deformation of Composite Materials, Proc. IUTAM Symp.*, Troy, New York, 1990, Ed. G.J Dvorak 193–214. Springer, New York.

Schapery, R. A. 1993. Viscoelastic deformation behavior of ice based on micromechanical models. *Ice Mechanics, ASME, AMD.* New York, vol. 163:15–34.

Schapery, R. A. 1996. Characterization of nonlinear, time-dependent polymers and polymeric composites for durability analysis. *Proc. Int. Conf. Prog. Durability Anal. Composite Syst.*, Brussels, July 1995. Balkema, Rotterdam.

Schapery, R. A. 1997a. Linear elastic and viscoelastic deformation behavior of ice. *J. Cold Regions Eng.*, ASCE, 11(4):271–289.

Schapery, R. A. 1997b. Nonlinear viscoelastic and viscoplastic constitutive equations based on thermodynamics. *Mech. Time-dependent Mat.* 1:209–240

Schulson, E. M. 1999. The structure and mechanical behavior of ice. *JOM*, 51(2), pp. 21–27.

Schulson, E. M. and Duval, P. 2009. *Creep and Fracture of Ice.* Cambridge University Press

Sibson, R. H. 1977. Fault rocks and fault mechanisms. *J. Geol. Soc. Lond.*, 133:191–213

Simonson E. R., Jones A. H. and Green S. J. 1975. High pressure mechanical properties of three frozen materials. *Fourth Int. Conf. High Pressure*, Kyoto, International Association for the Advancement of High Pressure Science and Technology, Physico-Chemical Society of Japan, Kyoto, 115–121.

Singh, S. K. 1993. Mechanical Behaviour of Viscoelastic Material with Microstructure. Ph.D. Thesis, Memorial University of Newfoundland, St. John's, Canada.

Singh, S. K., Jordaan, I. J., Xiao, J. and Spencer, P. A. 1995. The flow properties of crushed ice, *Proc. 12th Int. Conf. Offshore Mech. Arctic Eng. (OMAE)*, Glasgow, U. K. Journal of Offshore Mechanics and Arctic Engineering, vol. 117, 1995, 276–282.

Singh, S. K. and Jordaan, I. J. 1996. Triaxial tests on crushed ice, *Cold Reg. Sci. Technol.*, vol. 24, 153–165.

Singh, S. K. and Jordaan, I. J. 1999. Constitutive behaviour of crushed ice. *Proc. Symp. Inelasticity and Damage in Solids subject to Microstructural Change*, Memorial University of Newfoundland, 1997: 367–379. Extended version in Int. J. Fract. 1999;97(1–4):171–187.

Sinha, N. K. 1977. Technique for studying the structure of sea ice. *J. Glaciol.* 18, 315–323.

Sinha, N. K. 1978. Short-term rheology of polycrystalline ice. *J. Glaciol.*, 21(85):457–473.

Sinha, N. K. 1979. Grain boundary sliding in polycrystalline materials. *Phil. Mag.* 40(6): 825–842.

Sinha, N. K. 1984. Uniaxial compressive strength of first-year and multi-year sea ice. *Can. J.Civ. Eng.* 11. 82–91.

Sinha, N. K. 1988. Crack-enhanced creep in polycrystalline material: strain-rate sensitive strength and deformation of ice. *J. Mater. Sci.*, 23:12, 4415–28.

Sinha, N. K. 1991. Kinetics of microcracking and dilatation in polycrystalline ice. *Ice-Structure Interaction. Proc. IUTAM/IAHR Symp.*, St John's, Canada, 1989. Springer-Verlag, Berlin, 69–87.

Sinha, N. K., and Cai, B. 1992. *Analysis of Ice from Medium-Scale Indentation Tests.* NRC Laboratory Memorandum IME-CRE-LM-002.

Sjölind, S-G. 1987. A constitutive model for ice as a damaging viscoelastic material. *Cold Reg. Sci. Technol.* 14,3:247–262.

Sneddon. I. N. 1964. Technical Report AFOSR 64–1989, North Carolina State University.

Spencer, P. A., Masterson, D. M., Lucas, J. and Jordaan, I. J. 1992 The flow properties of crushed ice 1: experimental observation and apparatus. *Proc. 11th IAHR Int. Ice Symp.*, Banff, Alberta, vol. I, 258–268.

St. John, J. W., Daley, C., Blount, H. and Glen, I. 1984. *Ice Loads and Ship Response to Ice USCG Polar class Deployment.* Technical report prepared for the Transport Canada.

St. John, J. and Minnick, P., 1993. *Swedish Icebreaker Oden Ice Impact Load Measurements during International Arctic Ocean Expedition 1991; Instrumentation and Measurement Summary,* STC Tech. Rep. 2682 to U.S Coast Guard Headquarters, May.

Stone, B. M., Jordaan, I. J., Jones, S. J. and McKenna, R. F. 1989. Damage of isotropic polycrystalline ice under moderate confining pressures, *Proc. 10th. Int. Conf. Port Ocean Eng. Arctic Cond. (POAC),* Lulea, Sweden, June, vol. 1, 408–419.

Stone, B. M., Jordaan, I. J., Xiao, J. and Jones, S. J. 1997. Experiments on the damage process in ice under compressive states of stress, 1996, *J. Glaciol.* 43(143):11–25.

Strecker, K., Ribeiro, S. and Hoffmann, M.-J. 2005. Fracture toughness measurements of LPS-SiC: a comparison of the indentation technique and the SEVNB method. *Mater. Res.,* 8(2):121–124.

Stuckey, P., Ralph, F. and Jordaan, I. 2008. Iceberg design load methodology. *Proc. 8th Int. Conf. Perform. Ships Structures Ice (ICETECH 2008).* Banff, Canada, 2008.

Stuckey, P and Fuglem, M. 2014. Challenges in determining design iceberg impact loads for offshore structures. *Proc. ICETECH 2014,* Banff, AB, Canada.

Takeuchi, T., Masaki, T., Akagawa, S., et al., 1997. Medium-scale indentation tests (MSFIT) - ice failure characteristics in ice–structure interactions. *Proc. 7th Int. Offshore Polar Eng Conf,* Honolulu, Hawaii, ISOPE, vol. 2. 376.

Takeuchi, T., Akagawa, S., Kawamura, M., et al., 2000. Examination of factors affecting total ice load using medium field indentation test data. *Proc. 10th Int. Offshore Polar Eng. Conf.,* Seattle, Washington, ISOPE, vol. 2, 607.

Taylor, R. S. 2010. Analysis of Scale Effect in Compressive Ice Failure and Implications for Design. PhD Thesis. Memorial University.

Taylor, R. S., Jordaan, I. J., Li, C. and Sudom, D. 2010. Local design pressures for structures in ice. Analysis of full-scale data. *J. Offshore Mech. Arctic Eng., ASME,* August;132(3), 031502-1-7.

Taylor, R. S. and Jordaan, I. J. 2011a. The Effects of non-simultaneous failure, pressure correlation, and probabilistic averaging on global ice load estimates. *Proc. 21st Int. Offshore Polar Eng. Conf.,* Maui, Hawaii, USA, June 19–24.

Taylor, R. S. and Jordaan, I. J. 2011b. Pressure-area relationships in compressive ice failure: application to Molikpaq. *Proc. 21st Int. Conf. Port Ocean Eng. Arctic Cond. (POAC),* July 10–14, Montréal, Canada. Paper no. POAC11-158.

Taylor, R. S. and Jordaan, I. J. 2015. Probabilistic fracture mechanics analysis of spalling during edge indentation in ice. *Eng. Fract. Mech.* 134:242–266.

Timco G. W. and Frederking R. M. W. 1986. The effects of anisotropy and microcracks on the fracture toughness of freshwater ice. *Proc. 5th Int. Offshore Mech. Arctiv Eng. (OMAE).* Tokyo, 1986.

Tunik, A. L. 1991. Impact ice pressure: more questions than answers. *Ice–Structure Interaction. Proc. IUTAM/IAHR Symposium,* St John's, Canada, 1989. Springer-Verlag, Berlin, 693–714.

Turner, J. 2018. Constitutive Behaviour of Ice Under Compressive States of Stress and its Application to Ice-Structure Interactions. PhD thesis, Memorial University of Newfoundland.

Urabe, N., Iwasaki, T. and Yoshitake, A. (1980). Fracture toughness of sea ice. *Cold Reg. Sci. Technol.* 3:29–37.

Urai, J. L., Means, W. D. and Lister, G. S. 1986. Dynamic recrystallization of minerals. Mineral and Rock Deformation: Laboratory Studies: The Paterson Volume. *American Geophysical Union*, 161–199.

Vanmarcke, E. 1983. *Random Fields: Analysis and Synthesis*. MIT Press, Cambridge Mass.

Wang, Y. S. 1979. Crystallographic studies and strength tests of field ice in the Alaskan Beaufort Sea. *Proc. 5th Int. Conf. Port Ocean Eng. Arctic Cond. (POAC'79)*, 651–655.

Wang, Y. S. and Poplin, J. P. 1986. Laboratory compressive tests of sea ice at slow strain rates from a field test program. *Proc. 5th Int. Offshore Mech. Arctic Eng. (OMAE 86)*, Tokyo, vol. 4, 379–384.

Weeks, W. F. 2010. *On Sea Ice*. University of Alaska Press.

Weertman, J. 1969. Effects of cracks on creep rate. *Q. Trans. ASM*, 62(2): 502–511.

Weibull, W. 1939. *A Statistical Theory Of The Strength Of Materials*. Ingeniörsvetenskapsakademiens Handlingar Nr 151, 1939, Generalstabens Litografiska Anstalts Förlag, Stockholm.

Weibull, W. 1951. A statistical distribution function of wide applicability. *J. Appl. Mech.*, 18.

Weiss, J. and Schulson, E. M. 1995. The failure of fresh-water granular ice under multiaxial compressive loading. *Acta Metal. Mat.* 43, 2303–2315.

Wells, J., Jordaan, I., Derradji-Aouat, A. and Taylor, R. 2011. Small-scale laboratory experiments on the indentation failure of polycrystalline ice in compression: main results and pressure distribution. *Cold Reg. Sci. Technol.* 65 314–325.

Widianto, W., Khalifa, J., Younan, A., Karlsson, T., Stuckey, P. and Gjorven, A. 2013. Design of Hebron gravity based structure for iceberg impact. *Proc. 23rd Int. Offshore Polar Eng. Conf.*, vol. 1. Anchorage, Alaska, USA, 2013.

Wilkinson, D. S. and Ashby, M. F. 1975. Pressure sintering by power-law creep. *Acta Metal.* 23, 1277–1285.

Williams, M. L. 1957. On the stress distribution at the base of a stationary crack. *J. Appl. Mech.*, 24:109–114.

Woytowich, R. 2003. Riveted hull joint design in RMS *Titanic* and other pre–World War I ships. *Mar. Tech.* 40 (2):82–92

Xiao, J. (Jing). 1991. Finite Element Modelling of Damage Processes in Ice-Structure Interaction. M.Eng Thesis, Memorial University of Newfoundland, 98.

Xiao, J., Jordaan, I. J., McKenna, R. F. and Frederking, R. M. W. 1991. Finite element modelling of spherical indentation tests on ice. *Proc. 11th. Int. Conf. Port Ocean Eng. Arctic Cond.*, vol. 1, St. John's, September 24–28, 471–485.

Xiao, J., Jordaan, I. J. and Singh, S. K. 1992. Pressure melting and friction in ice–structure interaction. *Proc. 11th IAHR Int. Ice Symp.*, Banff, Alberta, vol. 3, 1255–1268.

Xiao, J. and Jordaan, I. J. 1996. Application of damage mechanics to ice failure in compression. *Cold Reg. Sci. Technol.* 24:305–322.

Xiao, J. (Jing) 1997. Damage and Fracture of Brittle Viscoelastic Solids with Application to Ice Load Models. Doctoral thesis, Memorial University of Newfoundland, 187.

Zou, B., Xiao, J. and Jordaan, I. J. 1996. Ice fracture and spalling in ice-structure interaction. *Cold Reg. Sci. Technol.* 24 (2): 213–220.

Zou, B (Bin). 1996. Ships in Ice: the Interaction Process and Principles of Design. Doctoral thesis, Memorial University of Newfoundland, 181.

Index

activation energy, 25
activation energy in triaxial tests, 128
alpha method, 35
analysis of indentation with full damage
    model, 139
Andrade relationship, 18
anisotropy, 6
area definitions for local and global
    design, 31

Baltic Sea, 7, 69
Barenblatt crack, 171
basal plane
    basal plane, 5
basal slip, 17
beam bending fracture tests, 187
Beaufort Sea, 4, 41, 52
bergy bit, 9, 42
Biot-Schapery theory, 193
bow form, 44
breakdown of structure in triaxial tests, 123
brine, 9, 10
Burgers model, 16, 95, 100, 118, 145,
    152, 153
Burgers model with damage, 135
Burgers model with stress-dependent
    dashpots, 114

c-axis, 5, 10, 22, 209
Canada, 4
cataclasis, 14
cavities in crushed ice analysis, 152
climate, 6
codes and standards, 80
columnar ice, 9, 132, 170
compact tension fracture tests, 188
comparison of fracture test results and theory, 189
components of creep, 16
compound Poisson process, 39
confining pressure, 20, 112, 115, 116, 120,
    128, 209
constitutive theory and MSP, 199
correspondence principle (Schapery), 201
correspondence principle and MSP, 111

correspondence principle in linear
    viscoelasticity, 103
Coulomb-Mohr flow, 149
crack, 13, 23, 26, 65, 103, 104, 115, 117, 122, 124,
    151, 165, 169, 171, 177, 182, 187, 201, 202
crack density, 104
crack growth, 181, 204, 205
crack opening displacement, 179
crack-enhanced creep, 105
creep, 12, 15, 75, 86, 100, 105, 114
crushed ice constitutive behaviour, 151
crushed layer as linear viscous fluid, 91
crushing, 13

damage, 13, 103, 109, 117, 135, 151, 202
damage analysis with pressure softening, 135
damage effect in uniaxial tests, 105
damage effect on triaxial response, 122
damage effect with low confinement, 107
damage formation and indentation speed, 142
damage formulation as a function of
    pressure, 139
damage localization in triaxial tests, 130
damage mechanics, 103, 135, 202
damage theoretical formulation, 206
damage theory and J integral, 202
damaged layer at different scales, 167
dashpot, 3, 87, 91, 114, 136
deformation
    and microstructure, 116
    components of, 15
    crack opening, 178
    field ice, 142
    ice, 12, 87
    localization of, 103
    mechanism, 21
    Molikpaq extensometers, 54
    recrystallization, 20
    thermally activated, 24
    time-dependent, 12, 85
    viscoelastic, 68
design for ice loading, 71
design for local pressures, 43
dissipation, 3

dropped ball tests, 91
dynamic recrystallization, 20

entropy, 15
    hydrogen bonds, 6
    melting, 19
    production coefficient, 195
    production in irreversible process, 194
exposure, 33
extrusion of crushed ice, 148

failure modes in triaxial tests, 121
failure zone size, 182
finite element analysis, 117
finite element analysis of field indentation tests,
    119
first year ice, 8
flexible indentors, 160
fluctuations in load, 145
fracture, 13, 25, 169
fracture and damage in indentation, 157
fracture energy, 178
fracture of ice and time dependency, 169
fracture toughness, 26, 170, 187
freezing process, 7

geophysical analogies, 14
global force and high-pressure zones, 76
global ice pressures, 75
global pressure-area relationships, 75
grain boundary sliding, 17

Hans Island, 57
Helsinki tests, 47
hereditary integrals, 95
Hobson's Choice tests, 60
Huygens' clocks, 163

ice compressive strength, 10
ice crystal, 4
ice failure, 10
ice management, 74
ice on the earth, 6
ice-induced vibrations, 68
iceberg, 9
iceberg field tests, 70
IDA project, 40
imperfections and runaway strains, 134
incompressible power-Law materials, 146
incremental creep analysis, 137
indentation and temperature, 157
ISO 19906, 72

J integral, 110, 182, 202
JOIA indentor tests, 48

Kelvin model, 93, 137, 140
Kigoriak, 34

laboratory test details, 208
landfast ice, 75
layer, 11
layer development, 21
limit states, 72
linear viscoelastic analsis of fracture, 177
linear viscoelasticity, 87
local and global pressure, 30
local ice pressure, 31, 34, 37, 42, 55, 61, 73, 80
localization, 11, 21, 130, 167
localization and stress concentrations, 120
localization in indentation tests, 120
lock-in, 160
Louis S. St-Laurent, 37, 40
lubrication theory, 92

Maxwell model, 87, 146
mechanics, 14
medium scale indentation tests, 59
Medof panels, 54
melting during indentaion, 154
microracking damage, 104
modified superposition principle (MSP), 108
Molikpaq, 52
Monte Carlo method, 73
MSP with nonlinearity and damage, 109
multiaxial states of stress, 112
multiyear field ice and damage state, 118
multiyear ice, 9

Newtonian fluid, 91
nonlinear dashpots, 100
nonlinear viscoelastic fracture theory, 182
Northern Sea Route, 74

Oden, 41

plasticity, 85, 116, 182
plasticity theory, 85
platen effects, 115
Poisson process, 33
Polar Sea, 40
Pond Inlet Tests, 59
power-law potentials, 147
pressure hardening and softening, 117, 136
pressure melting, 18
pressure softening, 119
pressure-area relationship, 30
pressures on the Molikpaq, 55
primary, secondary and tertiary creep, 15
probabilistic averaging, 51
probability-based design, 72

Rae Point tests, 60
rare and frequent loads, 72
redistribution of stress and strain, 102
relaxation spectrum, 100
retardation spectrum, 98
runaway strains, 131

sawtooth loading, 160
scaling in ice testing, 164
ship trials, 34
sintering, 154
small scale indentation tests, 156
spectrum for Andrade relationship, 101
spring, 3, 87
springs and dashpots, 87, 198
strength, 13
stresses and flux of energy in analysis of
    indentation, 142
STRICE, 49
synchronization of high-pressure zones, 163

targeted triaxial tests, 119
temperature, 157
temperature and indentation analysis, 143
temperature fluctuations in indentation tests, 154
Terry Fox, 42
theoretical strength, 26
thermally activated processes, 24
thermodynamics, 15

thermodynamics of irreversible processes, 193
time dependency of fracture, 27, 177
Titanic, 79
triaxial test review, 115
true triaxial testing, 115

velocity effects, 44
viscoelastic theory, 85
viscous behaviour and Coulomb-Mohr assumption
    for crushed ice, 149
von Mises stress and strain, 113
von Mises yield criterion, 114

weakest link theory, 166
Weibull theory, 166
work of R. A. Schapery, 192

yield stress ambiguities, 85
young ice, 8

zonal forces, 38